About Island Press

Since 1984, the nonprofit Island Press has been stimulating, shaping, and communicating the ideas that are essential for solving environmental problems worldwide. With more than 800 titles in print and some 40 new releases each year, we are the nation's leading publisher on environmental issues. We identify innovative thinkers and emerging trends in the environmental field. We work with world-renowned experts and authors to develop cross-disciplinary solutions to environmental challenges.

Island Press designs and implements coordinated book publication campaigns in order to communicate our critical messages in print, in person, and online using the latest technologies, programs, and the media. Our goal: to reach targeted audiences—scientists, policymakers, environmental advocates, the media, and concerned citizens—who can and will take action to protect the plants and animals that enrich our world, the ecosystems we need to survive, the water we drink, and the air we breathe.

Island Press gratefully acknowledges the support of its work by the Agua Fund, Inc., The Margaret A. Cargill Foundation, Betsy and Jesse Fink Foundation, The William and Flora Hewlett Foundation, The Kresge Foundation, The Forrest and Frances Lattner Foundation, The Andrew W. Mellon Foundation, The Curtis and Edith Munson Foundation, The Overbrook Foundation, The David and Lucile Packard Foundation, The Summit Foundation, Trust for Architectural Easements, The Winslow Foundation, and other generous donors.

The opinions expressed in this book are those of the author(s) and do not necessarily reflect the views of our donors.

RESTORING DISTURBED LANDSCAPES

River Futures: An Integrative Scientific Approach to River Repair,
edited by Gary J. Brierley and Kirstie A. Fryirs

Large-Scale Ecosystem Restoration: Five Case Studies from the United States, edited by Mary Doyle and Cynthia A. Drew

New Models for Ecosystem Dynamics and Restoration,
edited by Richard J. Hobbs, and Katharine N. Suding

Cork Oak Woodlands in Transition: Ecology, Adaptive Management, and Restoration of an Ancient Mediterranean Ecosystem, edited by James Aronson, João S. Pereira, and Juli G. Pausas

Restoring Wildlife: Ecological Concepts and Practical Applications, by Michael L. Morrison

Restoring Ecological Health to Your Land,
by Steven I. Apfelbaum and Alan W. Haney

Restoring Disturbed Landscapes: Putting Principles into Practice,
by David J. Tongway and John A. Ludwig

SOCIETY FOR ECOLOGICAL RESTORATION INTERNATIONAL

The Society for Ecological Restoration (SER) is an international nonprofit organization whose mission is to promote ecological restoration as a means to sustaining the diversity of life on Earth and reestablishing an ecologically healthy relationship between nature and culture. Since its foundation in 1988, SER has been promoting the science and practice of ecological restoration around the world through its publications, conferences, and chapters.

SER is a rapidly growing community of restoration ecologists and ecological restoration practitioners dedicated to developing science-based restoration practices around the globe. With members in more than 48 countries and all 50 U.S. states, SER is the world's leading restoration organization. If you wish to become a member, contact SER at 285 W. 18th Street, #1, Tucson, AZ 85701. Tel. (520) 622-5485, e-mail: info@ser.org. www.ser.org.

The opinions expressed in this book are those of the authors and are not necessarily those of SER.

Restoring Disturbed Landscapes

Putting Principles into Practice

David J. Tongway and John A. Ludwig

SOCIETY FOR ECOLOGICAL RESTORATION INTERNATIONAL

Washington | Covelo | London

Library of Congress Cataloging-in-Publication Data

Tongway, David J. (David John)
  Restoring disturbed landscapes : putting principles into practice / David J. Tongway and John A. Ludwig.
    p. cm. — (The science and practice of ecological restoration)
  Includes bibliographical references and index.
  ISBN-13: 978-1-59726-580-5 (cloth : alk. paper)
  ISBN-10: 1-59726-580-2 (cloth : alk. paper)
  ISBN-13: 978-1-59726-581-2 (pbk. : alk. paper)
  ISBN-10: 1-59726-581-0 (pbk. : alk. paper) 1. Restoration ecology. 2. Landscape protection. 3. Landscape ecology. I. Ludwig, John A. II. Title.
  QH541.15.R45T66 2010
  639.9—dc22
                            2010013877

Printed on recycled, acid-free paper ♲

Manufactured in the United States of America
10 9 8 7 6 5 4 3 2 1

The information in this book is accurate to the best of the authors' knowledge. However, neither Island Press nor the authors are responsible for injuries that may occur as a result of the restoration procedures or activities described in this book.

Keywords: disturbed landscape, functional landscape, dysfunctional landscape, restoration, landscape restoration, restoration practitioner, landscape function analysis, landscape ecology, ecological restoration, open-cut coal mining, mine-site tailings storage, mine-site waste-rock dump, road verge restoration, tree belt, water ponding, soil-surface indicator, adaptive management, monitoring indicator

To our wives, Helen Tongway and Rosalind Ludwig,
who encouraged us in so many, many ways to write this book.

*A complex system that works is invariably found to have
evolved from a simple system that works.*

John Gall

CONTENTS

FOREWORD James Aronson   xv

PREFACE   xix

ACKNOWLEDGMENTS   xxi

PART I A Function-Based Approach to
Restoring Disturbed Landscapes   1

Chapter 1 Our Approach to Restoring Disturbed Landscapes:
Five-Step Adaptive Procedure   3

Chapter 2 A Framework for How Landscapes Function   7

Chapter 3 Principles for Restoring Landscape Functionality   19

PART II Case Studies on Restoring Landscapes: Mine Sites
and Rangelands   27

Chapter 4 Restoring Mined Landscapes   29

Chapter 5 Restoring Damaged Rangelands   45

PART III Scenarios for Restoring Landscapes: Mine Sites,
Rangelands, Farmlands, and Roadsides   63

Chapter 6 Restoration of Mine-Site Waste-Rock Dumps   65

Chapter 7 Restoration of Mine-Site Tailings Storage Facilities   75

Chapter 8 Restoring Landscapes after Open-Cut Coal Mining   87

Chapter 9 Restoring Rangelands with an Overabundance of Shrubs   97

Chapter 10 Renewing Pastureland Functions Using Tree Belts   107

Chapter 11 Restoration of Former Farmlands near
Urban Developments   117

Chapter 12 Restoring Verges after Road Construction   129

PART IV Monitoring Indicators   137

    Chapter 13 Landscape Function Analysis: An Overview and
                 Landscape Organization Indicators   139

    Chapter 14 Landscape Function Analysis: Soil-Surface Indicators   145

    Chapter 15 Ephemeral Drainage-Line Assessments:
                 Indicators of Stability   151

    Chapter 16 Vegetation Assessments: Structure and Habitat
                 Complexity Indicators   157

    Chapter 17 Reflections on Restoring Landscapes:
                 A Function-Based Adaptive Approach   163

REFERENCES   167

GLOSSARY   173

FURTHER READING   179

ABOUT THE AUTHORS   181

INDEX   183

FOREWORD

Building on thirty-five years of collaborative reflection and observation, plus a vast amount of fieldwork and teaching, David Tongway and John Ludwig have written a marvelous book summing up their intertwined and complementary stocks of knowledge and experience in the science and practice of ecological restoration. This renowned duo have devoted their professional lives to understanding how natural landscapes work as biophysical systems, and how they are damaged by disturbances of various kinds. The assumption—borne out by much testing in the field—is that ecosystem rehabilitation and restoration strategies to some specified goal can become a much more assured process when based on the principles identified in the functional analysis of landscapes, and when applying a rational, stepwise approach, with regular reference to a specific, well-studied situation.

The approach and procedures provided here will help you learn not only how to "read" almost any terrestrial landscape, in a detailed but highly practical way, but also how to use that skill set for purposes of designing workable solutions to repair damage for your specific case, without getting locked into trying to apply restoration "recipes" or off-the-shelf approaches that very likely may not work. As the authors remind us, "you can't fix something unless you know how it works" (part 1). Focusing on biophysical realities at the landscape scale, *Restoring Disturbed Landscapes* presents a readily transferable approach and illustrates how restoration practitioners can reverse declines in the flow of goods and services supplied by landscapes of all kinds, provided they have studied how the systems work in the first place.

What the authors present, in a lucid, fluid, and accessible style, supported by ample illustrations and examples from their own experience, is their widely tested five-step procedure—setting clear goals; defining the problem; designing solutions; applying technologies and monitoring their effects; and, finally, as needed, adaptively improving technologies. Although first developed in arid and semiarid lands in Australia, Tongway and Ludwig's procedure and the underlying conceptual model, which the reader will find in chapter 2, is applicable in high rainfall areas as well. This model provides the framework for the five-step procedure that has been tested by the authors in dozens of difficult sites and imparted to dozens of student and trainee groups around the world.

I am convinced that for those practitioners who discover these authors for the first time, as well as for all those who have had the good fortune to participate in Dave Tongway's training sessions in landscape function analysis procedures, this book will be invaluable. It makes available the sound scientific underpinnings of landscape function analysis, stated in a manner that laypeople and students can follow. The authors wrote to me recently that "having observed a lot of poor rehabilitation and ineffective monitoring, we wanted to provide a synthesis of the needs for both activities, but not in a site-specific manner. This is where our articulation of the principles is so potentially useful to folk who want to do a good job, but only have intuition and rumors to work with. We want to open up all the science needed for restoration, but in an uncomplicated way. Few ordinary folk will have had the opportunity of reading the scientific literature as broadly as we have, so the book is to show that science is useful at the grassroots level of application."

This book will be of great interest to, and use for, teachers, students, volunteer and professional restoration practitioners, and, indeed, a wide range of scientists more or less closely engaged with restoration, as well. As anyone familiar with arid and semiarid land ecology, in particular, will recognize, the late, eminent ecologist Immanuel Noy-Meir was a strong influence on Dave and John, by drawing their attention to the need for heterogeneity of water distribution in semiarid and arid landscapes. A finer pedigree than that, one could hardly ask for. And Noy-Meir himself enthusiastically acknowledged what an important contribution the team of John Ludwig and David Tongway has made to applied ecology.

A word about The Science and Practice of Ecological Restoration Series: since the first volume appeared in 2002, this series has sought to embrace and, indeed, illuminate the entire breadth of the rapidly evolving field of ecological restoration, including all the different sciences and all the different forms of practice. The range of books nicely reflects SER and Island Press's shared mission vis-à-vis an ever-wider public concerned with conservation, sustainable use of resources, sustainability in general, and ecological restoration. However, there is certainly a need for more books that primarily address practitioners—and beginning students. This fine work by Tongway and Ludwig, which is the twentieth volume to appear in the series, goes a long way to rectify the balance, and it has the great virtue of being written by people who bridge the divide between scientists and practitioners. The scientific credentials and underpinnings of Tongway and Ludwig are as impeccable as their field experience—which is a(nother) rare and valuable asset, although it is a hallmark of this series that many of the contributing authors help make this vital bridge between science and effective practice. That way lies sustainability and restoration not only of landscapes but also of societies, cultures, and a global community looking for new models and alternative futures.

Global society urgently needs to change the way we do many things if we wish for a sustainable and desirable future. We need to control population growth; heavily tax stock market speculators; remove perverse subsidies wherever they persist; reduce ridiculous consumption patterns among the affluent and the very afflu-

ent; and deeply respect the rights of all cultures, all species, and future genera-
tions of people. In light of the sorry state of our environment and ecosystems, and
our overall relationship with the natural world, we also need to start investing much
more heavily in the science, practice, and teaching of ecological restoration.

"Being able to 'read the landscape' is a rare and valuable asset." Those are the
last words of this book you are about to read, and it won't spoil anything to quote
them here. If you want to become more effective in restoring the functional capacity of disturbed landscapes and damaged ecosystems around the globe, you need
to learn how to read landscapes, so please read and ponder this valuable and much-
awaited book. Then, if you agree it is useful, share it with other people looking
for insight and inspiration from two of the most experienced restoration scien-
tists/practitioners anywhere.

James Aronson, editor
The Science and Practice of Restoration Ecology Series

In our changing world, human populations are rapidly growing and demanding more of the goods and services provided by landscapes. Restoration practitioners are needed who can improve the capacity of damaged landscapes to carry on these functions. Our aim in writing this book is to provide these practitioners with an approach to restoring the functional capacity of landscapes. Our function-based approach centers on a five-step, adaptive landscape restoration procedure, which we have found effective in restoring the functional capacity of disturbed landscapes in a wide variety of environments around the globe. This approach builds on a solid foundation of ecological concepts and principles, which can readily be put into practice.

The level of functional improvement aimed for depends on the goals of stakeholders—those people with an interest in, or dependence on, a landscape, whether or not they live in that landscape. Stakeholders may wish to restore a landscape to a more natural state to achieve improved biodiversity goals, for example, and this book can help practitioners work with stakeholders to achieve such goals.

However, *Restoring Disturbed Landscapes* is not about returning damaged lands to some notional "pristine" state; it is about repairing landscapes to an acceptable level of functionality. Unfortunately, the terms *landscape renewal*, *reclamation*, and *rehabilitation* are frequently used to imply that a pristine state is aimed for, especially in cases where the goal is to restore habitats for specific fauna. In this book we use such terms in the broader sense of repairing landscapes, damaged by various land uses, to an agreed state, rather than the narrow definition of restoring lands to some notionally prior state.

At the heart of *Restoring Disturbed Landscapes* is the five-step procedure of setting clear goals, defining the problem, designing solutions, applying technologies, and monitoring their effects, and, if needed, adaptively improving technologies. We will explain the principles behind this adaptive landscape restoration procedure and present examples to demonstrate why we believe that putting these principles into practice leads to successful landscape restoration.

Restoring Disturbed Landscapes will be of interest to restoration practitioners, such as natural resource managers, mine-site rehab professionals, elected leaders responsible for public lands, scientists, educators training their students, and members of the public caring for their lands. We feel that these dedicated people need easy-to-understand information that explains how to put basic principles into practice to facilitate the attainment of their landscape-restoration goals. This need was confirmed by practitioners who participated in workshops held around Australia and who contributed to the report "Restoring landscapes with confidence" (Lovett et al. 2008). Our aim is to meet this need and to encourage practitioners to think broadly and critically about their restoration problems, so that they can achieve their goals with as few "wrong turns" as possible.

David J. Tongway
John A. Ludwig
June 2010

ACKNOWLEDGMENTS

We are indebted to our many colleagues from Australia's Commonwealth Scientific and Industrial Research Organisation (CSIRO) for helping us form, test, and solidify the concepts and principles described in this book, and in particular to Brian Walker who, as a former chief of our CSIRO Division, cut us enough slack to explore and develop ideas as an unplanned output of our research. The foundation for our conceptual framework on how landscapes function is based on the *pulse-and-reserve* framework for arid and semiarid systems presented in the 1980s in papers by Brian Walker, Professor Mark Westoby, and the late, eminent desert ecologist Immanuel Noy-Meir.

We acknowledge with affection and appreciation the mentoring role, to both of us, of Walt Whitford, who as a friend and ecologist at New Mexico State University opened our eyes to the breadth of biotic and abiotic interactions in landscapes.

We thank the many practitioners who provided us valuable feedback on ways to clearly communicate concepts and principles in landscape ecology. We also thank Elizabeth (Ludwig) Tomko of Tomko Design for suggesting possible front cover designs, Yvette Salt for designing figure 1.1 (chapter 1), and all those who contributed images to this book (as acknowledged in the figure captions) where our own collections were insufficient.

We are indebted to the Society of Ecological Restoration (SER) for accepting this book into their series of distinguished Island Press publications. In particular we appreciate the enthusiastic support of James Aronson from the earliest stages of our book proposal to reviewing early drafts and writing a foreword. Previous SER/Island Press titles edited by Aronson and Andre Clewell were valuable guides in providing definitions we have adopted as standards. We especially value the professionalism of Island Press editors Barbara Dean, Erin Johnson, and Sharis Simonian, whose prompt responses helped us maintain our writing momentum and greatly improved our book.

David J. Tongway
John A. Ludwig

PART I

A Function-Based Approach to Restoring Disturbed Landscapes

The world's landscapes provide goods, such as food, fiber, and minerals, and ecosystem services, such as clean water, to billions of people. The demand for such goods and services is increasing as human populations grow, but the capacity of landscapes to meet human needs is diminishing. Consequently, it is critical that we address this supply and demand problem before further declines in capacity occur. This book illustrates how restoration practitioners can reverse declines in goods and services by improving the functional capacity of damaged landscapes.

We begin by defining what we mean by *landscapes* and *functional landscapes*. Landscapes are areas of interconnected ecosystems, which are communities of organisms interacting with one another and their physical environment. Organisms are the biotic components of ecosystems and include plants (primary producers); animals (macroconsumers); and microorganisms (microconsumers), such as fungi and bacteria. Microorganisms are included because they decompose complex organic materials tied up in dead organisms and release these materials as simple mineral nutrients, which are then available for use again by plants and animals. Physical or abiotic components of ecosystems include, for example, mineral ions in soils, such as calcium, potassium, and phosphate, and climatic factors, such as precipitation and temperature. Humans are part of ecosystems and the greater landscape because we influence their functioning or health both directly and indirectly.

Functional landscapes are those that have a high capacity to provide important biophysical and socioeconomic goods and services (Ludwig et al. 1997; Tongway and Ludwig 2007). Landscape functions include the following:

- Maintaining basic processes such as capturing energy, retaining and using water, and cycling nutrients
- Providing habitats for populations of plants, animals, and microorganisms
- Sustaining people by providing their material, cultural, and spiritual needs

Dysfunctional landscapes have impaired capacities for one or more of these functions. Restoring disturbed landscapes essentially means repairing damaged functions.

We have organized part 1 into three chapters. In chapter 1 we describe our approach to restoring landscapes—an orderly five-step adaptive procedure. In chapter 2 we present a conceptual framework that we have found useful in helping us

1

to understand the complexities of how landscapes function or work as biophysical systems. Then in chapter 3 we build on this conceptual framework by defining four principles underlying adaptive landscape restoration. We believe that in order to be able to restore disturbed landscapes it is essential that practitioners understand landscape functioning concepts and principles. The well-known axiom "You can't fix something unless you know how it works" emphasizes this point. Throughout this book we refer to axioms when we find them pertinent.

In the chapters of part 2, we demonstrate how practitioners have put the concepts and principles of landscape restoration into practice by describing case studies. In part 3, we build on these case studies by outlining scenarios of successful landscape restoration from around the globe. In part 4, we provide practitioners with descriptions of the tools and methods they need to evaluate landscape dysfunction and to monitor restoration trends toward desired goals.

Chapter 1

Our Approach to Restoring Disturbed Landscapes: A Five-Step Adaptive Procedure

In this chapter we describe our approach, which we feel is central to *Restoring Disturbed Landscapes*. We think of our approach as an orderly five-step adaptive procedure for restoring landscapes, or for short, *adaptive landscape restoration*. It comprises a sequence of steps (figure 1.1) where stakeholders in the disturbed landscape work with the restoration practitioner (hereafter abbreviated as RP) to (1) articulate the goals of restoration, (2) define and carefully analyze the problem, (3) identify appropriate solutions, (4) select treatments to apply, and (5) monitor restoration indicators and assess progress as trends in these data. If trends are negative, RPs should then adaptively revise treatments to improve trends. In practice, we view this procedure like planning a journey using a road map. The start and end points are known, but there are options to exercise in the actual route taken to reach the destination—the successful restoration of a landscape.

Before describing each of the five steps in our adaptive landscape restoration procedure we make a few observations and note some of its key features. We emphasize that our adaptive procedure is not a prescription or recipe for RPs to follow like a global positioning system (GPS)–guided route from start to finish. As with planning a journey, choices made at each road junction along the way require critical analysis and careful consideration. Sometimes the choice of route is crucial; sometimes options are roughly equivalent. The shortest route is not necessarily the most appropriate. For example, on revegetated mine sites, RPs often sow (in suitable

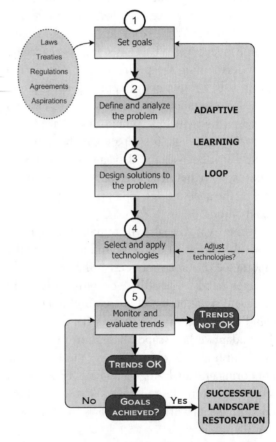

FIGURE 1.1. An orderly five-step procedure for restoring damaged landscapes that, if assessed trends are not OK, includes an adaptive learning loop to help achieve success by adjusting restoration technologies.

climates) the colorful, exotic red natal grass because it quickly provides an attractive cover. However, red natal is a tufted grass (big top, small base), and from our monitoring experiences, we know it usually fails to adequately function to prevent soil erosion on sloping sites (even if soil materials are only moderately erodible). We have found that a better option for protecting sloping rehabilitated surfaces against erosion is to revegetate mine sites with native, spreading perennial plants.

In designing ways to restore the functional capacity of damaged landscapes, we first need to be clear about what we are aiming to achieve. The next step is to critically analyze the problem: Which landscape goods and services have been lost and which have been retained? What landscape processes have become ineffective? What caused losses in capacity? By understanding the problem, RPs can design solutions by selecting and applying appropriate landscape restoration technologies. The restoration procedure must include monitoring, that is, RPs need to collect data to evaluate whether gains in capacity have been achieved. If gains have not been achieved, then they need to go back and reexamine whether the goals are still appropriate (e.g., are they too ambitious?).

In most cases, goals will remain appropriate, but we sometimes find that we underestimated the importance of some processes so that some technologies need to be adjusted. This revision to improve restoration is called an _adaptive learning loop_ (figure 1.1), which is an essential component of an adaptive landscape restoration procedure. After adjusting technologies, the RP must continue monitoring and analyzing data to see if restoration trends are toward desired goals. With time, these monitoring data will confirm if goals are being successfully achieved.

Our Five-step Procedure

Step 1: Setting goals
In step 1, stakeholders, or those with an interest in restoring a specified landscape, set goals that clearly define what they aim to achieve. When setting restoration goals, stakeholders require a clear under-standing of underlying constraints, such as whether goals are driven by regulations, laws, or treaties, or by agreements based on the aspirations of particular stakeholder groups. Often the groups involved in defining and setting restoration goals have different views about the initial state or condition of the landscape being restored, as well as having different desires for the shape, appearance, and final use of the restored landscape. To have clearly defined, agreed-upon goals, conflicting views and competing tensions need to be resolved.

Goals need to be stated in measurable terms so that RPs can collect the data that measures both progress toward the final goal and validates its achievement (i.e., whether rehabilitation trends are heading in the right direction and at an appropriate rate). If progress is lacking, monitoring data must provide the information needed for RPs to adjust technologies.

Step 2: Defining the problem
In step 2, stakeholders and RPs work together to carefully analyze the specific problem. This is a logical progression from having a clear understanding of constraints and goals. Step 2 involves a critical analysis of the landscape as a biophysical-socioeconomic system and the causes of the problem, not just a list of the symptoms. This analysis includes knowing the seriousness of the problem, how quickly it needs to be solved, and what information is available, or needs to be collected, to better understand the problem.

As well as identifying the underlying causes of the disturbance, we view step 2 as an analysis of damaged landscape processes, what we call _dysfunctional landscapes_ (Tongway and Ludwig 1996, 2007). Background information also needs to be evaluated, ideally from reference sites in landscapes that reflect the level of functionality aimed for in restoration goals. Reference sites provide stakeholders and RPs with a clearer perception of the magnitude of the problem by highlighting the gap in functionality between the damaged landscape and the reference landscape.

Step 3: Designing solutions

In the third step, stakeholders and RPs examine possible solutions to the problem. Their aim is to identify the biophysical, social, and economic processes that need to be improved to achieve the desired goals. We view step 3 as landscape restoration design, rather than technology selection, because to deal most effectively with the problem this step focuses on understanding and articulating the processes and functions that need restoring.

Step 4: Applying technologies

In step 4, stakeholders and RPs select appropriate solutions and technologies to apply. This is crucial, because if an inappropriate technology is chosen, or if an action is delayed, subsequent remedial (adaptive) actions may be very costly. For example, applying a treatment that inadvertently exposes a dispersive soil B horizon can lead to gully erosion that is very difficult and costly to repair.

Step 4 involves an *if–then* decision-making process where possible technologies are examined relative to problems and goals. Decision-making criteria include factors such as cost and technical difficulty. If, after being selected and applied, the RP finds that the technology does not lead to the desired goals, it is modified or replaced (revised by adaptive learning; figure 1.1). Examples of adaptive learning being applied to achieve desired restoration goals are presented in chapters of parts 2 and 3.

Step 5: Monitoring and assessing trends

The fifth step is about RPs monitoring the outcomes of applied technologies over time and then evaluating trends in monitoring data. It involves establishing a baseline by collecting data before, and as soon as possible after, implementing the selected rehabilitation technologies. To provide context and benchmark data, the RP needs to collect monitoring data from reference sites. These data provide the basis for evaluating the overall trend in rehabilitation progress over time and answering questions such as, are trends in the data toward values expected from measurements taken on reference sites?

We view the trend analysis phase of step 5 as an active and essential part of monitoring. This is important because if the RP detects unsatisfactory progress toward goals early on, rehabilitation technologies can be adapted to improve progress; this can greatly reduce costs of repairing future failures. The process of evaluating trends also adds to our knowledge about the landscape.

It is also important that RPs collect data over a sufficient length of time to detect genuine restoration trends. Then they can confidently ask the question, are trends OK? If not, this flags a need for RPs to reexamine goals, reanalyze the problem, and redesign solutions, which may simply lead to a modification of an already applied technology.

Further Thoughts

Over the years we have observed that less successful rehabilitation has usually glossed over one or more of the steps in our five-step adaptive landscape restoration procedure or has not put basic landscape function concepts and principles into practice. We describe these concepts and principles in chapters 2 and 3. We have observed that successful restoration can be achieved in deserts, grasslands, shrublands, savannas, woodlands, forests, and rainforests by putting these concepts and principles into practice within our five-step adaptive procedure. Sometimes this success was achieved by intuition, rather than by formally following our five-step procedure, but retrospective evaluation showed that all steps had been taken in order.

Chapter 2

A Framework for How Landscapes Function

People see landscapes differently: Some people respond mainly to the topography, that is, the shapes of mountains, hills, and valleys; others respond mainly to the diversity of the trees, shrubs, grasses, birds, and so on. However, to understand how disturbances affect landscapes so that a restoration practitioner (RP) can design effective projects, we believe it is important to view landscapes as functioning systems (Ludwig and Tongway 2000). We feel that with a function-based approach, landscape restoration becomes a matter of making the system work properly, rather than just replacing organisms that might be missing.

Landscapes, defined as interconnected ecosystems, are complex and dynamic (Tongway and Ludwig 2009). We marvel at the biological diversity of landscapes, such as tropical rainforests, and at the power of physical forces on landscapes, such as by glaciers. Landscapes that we also find fascinating are those that slowly tick along and then suddenly spurt into action because of an event such as a rainstorm. Arid and semiarid shrublands, grasslands, and savannas are examples of such landscapes; they have alternating slow and then fast dynamics, and they are sometimes referred to by scientists as *pulse-and-decline* and *stop-go* ecosystems, because they stop or slow down when water is limiting and become active when it is available (Westoby et al. 1989; Robin 2007).

Landscapes range in size (scale) from small to large depending on the extent of the area being considered by different groups. For example, a local community group may be interested in repairing a damaged hillslope of a few hectares, whereas a regional natural resource management group may be interested in the health of a river catchment of thousands of hectares. Given this variation in landscape scale, and the complexity and dynamics of landscapes, the question is, what can we do to help understand how disturbed landscapes function so that we can readily repair damaged processes?

In this chapter our aim is to describe what scientists call a *conceptual framework*, which is a device that enables us to structure or make sense of all the information we have about a landscape as a complex and dynamic process-based system (Ryan et al. 2007). In constructing a framework, we take care to be comprehensive and to not inadvertently omit important processes. Such frameworks, once fully developed, enable observers to read the landscape and understand why landscapes have become dysfunctional and fail, to some extent, to provide the goods and services desired by humans. We emphasize that conceptual frameworks help us design and apply effective restoration technologies by identifying defective or missing processes.

Although conceptual frameworks can take different forms, here we use box-and-arrow diagrams where we label boxes to represent the functioning biological (biotic) and physical (abiotic) components of the landscape system. We position boxes in our diagrams to represent the order, or sequence, of interactions between components of the system. Arrows are used to illustrate resource flows between and in and out of the components of the system. For example, rainwater enters a landscape as an external input and then flows as runoff within the landscape. These flows of water may then be captured by internal components, such as patches of vegetation, or water may run out of the defined landscape area into a different area, such as a creek or river located down the catchment. Flows of resources out of a landscape system represent external outputs.

Because water is essential to all life, the main focus in our conceptual framework is on the role of water. We provide examples of processes where water and materials such as soil particles and organic matter carried in runoff water enter, move around, are stored, utilized, and leave the landscape system. At times we also note how this framework applies to flows of materials driven by other forces, for example, wind-blown dust particles.

This focus of our conceptual framework on those biological and physical processes, such as water-driven soil erosion and growth of vegetation, is, in part, due to our area of expertise, but we feel strongly that this focus on biophysical processes is very important when restoring landscapes. We hasten to note, however, the important role that socioeconomic processes play in restoring landscapes, for example, in setting goals. (See step 1, figure 1.1 in chapter 1.) Socioeconomic criteria are also important in evaluating whether landscape restoration goals have been successfully achieved. Such evaluations can have huge economic consequences. For example, bonds worth millions of dollars may not be returned to mining companies after mine closure until they can demonstrate to regulators that they have successfully rehabilitated damaged landscapes.

Our Conceptual Framework

In this section we build a conceptual framework composed of landscape components (boxes) and processes (arrows). Our framework defines how materials flow into, around, and out of components of the landscape system. Our aim is to describe how landscapes function as complex and dynamic systems involving a sequence of important processes between components, but without introducing too much detail. We build our framework piece by piece, starting with external inputs and ending with external outputs, and put it all together in one diagram at the end. We illustrate how to put the framework to work by examples in chapters of parts 2 and 3. In the chapters of part 4, we describe how monitoring allows acquisition of data about changes and trends in landscape components and processes; these data are needed to evaluate whether trends are toward those specified by restoration goals.

Triggers and Transfers of Water

In pulse-and-decline landscape systems, dynamics are *triggered* into action, for example, by external inputs of water from rainfall events (figure 2.1a). We all know that storms can be intense and deliver large amounts of water in a short period of time over small areas. Intense storm events can trigger rapid and strong responses, including *transfer processes* (figure 2.1). In functional landscapes such events mostly involve redistribution of water within the landscape, but in disturbed landscapes disproportionate amounts flow out of the system as runoff (figure 2.1b). Functional landscapes have the capacity to capture high proportions of rainwater by the process of infiltration, because they typically have a high density and cover of vegetation that obstructs, contorts, and slows down overland flows (figure 2.1c). Dense vegetation allows more time for water to soak or infiltrate into the soil and for trapping the litter and the sediments and nutrients often carried in runoff.

These trigger and transfer processes also apply to small rainfall events, which can also produce infiltration and runoff but of smaller magnitude

Figure 2.1. Rainstorm events (a) can *trigger* large inputs of waters to a landscape, which then initiate *transfer* processes such as runoff and infiltration. A dysfunctional landscape (b) has a low capacity to capture flows of runoff water (illustrated by long, wide arrows representing flow pathways between vegetation patches and down the slope), whereas a functional landscape (c) has dense vegetation patches with a high capacity to obstruct runoff and to infiltrate water (depicted by numerous, short arrows representing contorted flow pathways).

than for intense events. Highly functional landscapes typically have the capacity to infiltrate most of the inputs of water during small, less intense events. In functional landscapes, small rainfall events are of great value to organisms, such as ephemeral plants (figure 2.2) or fungi. In dysfunctional landscapes, small rainfall events can quickly initiate runoff (figure 2.3). An example of this would be a landscape in which excessive grazing has reduced vegetation cover so that soil surfaces have become compacted and crusted. In this situation, flows are less impeded and surfaces readily erode (plate 2.1).

Reserves: Stores of Soil Moisture

In functional landscapes, processes such as the infiltration of rainwater into the soil will build up stores or *reserves* of moisture. Dysfunctional landscapes with low plant cover and soils with physical surface crusts and compacted subcrusts (figure 2.4a) have low infiltration rates, hence most inputs of rainwater flow or run off along soil surfaces causing erosion (figure 2.5a). Even on mildly sloping landscapes, runoff can become channelized into gullies (plate 2.2). Dysfunctional landscapes have a relatively low capacity

FIGURE 2.2. A highly functional landscape that retains water from small rainfall events, which is important for ephemeral plants (foreground).

FIGURE 2.3. An example of a dysfunctional landscape where water mostly runs off to trigger soil erosion.

(a)

(b)

FIGURE 2.4. (a) Soils with dense physical surface crusts and compacted subcrusts have low infiltration rates, which are typical of many dysfunctional landscapes around the globe. (b) Porous soils have high infiltration capacity, which typically occur under plants in highly functional landscapes.

to build reserves of soil moisture from given rain events.

Functional landscapes have healthy soils infused with plant roots and have a highly porous structure (figure 2.4b) and high infiltration rates. Functional landscapes have a high capacity for building

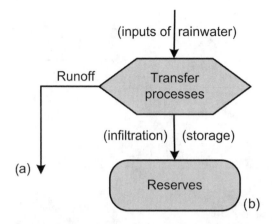

(a)

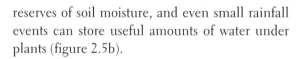

(b)

FIGURE 2.5. Inputs of rainwater to dysfunctional landscapes store little water because of runoff (a). Inputs of rainwater to functional landscapes initiate infiltration processes that store water in the soil, which can be called *reserves.* For example, an 11 mm input event wets a sandy loam soil to a depth of about 12 mm ([b] dark surface layer), and under the grass plant water flowed deeper along root channels or macropores ([b] dark tubes).

reserves of soil moisture, and even small rainfall events can store useful amounts of water under plants (figure 2.5b).

Pulses of Growth by Plants, Animals, and Microorganisms

If the amount of soil moisture stored in the soil is above that required by plants to initiate growth (an amount referred to as a *critical moisture threshold*), and other environmental factors, such as temperature, are favorable, these conditions can initiate responses or *pulses* of vegetative growth in a landscape (figure 2.6a). Microbial activity simultaneously leads to pulses of mineralization so that nutrients, such as nitrogen and phosphorus in plant-available forms, are produced. Thus plants grow, flower, and produce seeds (figure 2.6b). Also, by retaining vital water and nutrient resources in soil and plant reserves, functional landscapes provide the food and shelter (habitat) needs of wildlife (figure 2.7).

Gains to, and Losses from, Landscape Systems

To recap, inputs of water from rains trigger or drive transfer processes, such as infiltration, which in functional landscapes replenish soil water reserves,

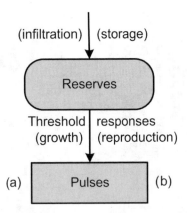

(a)

(b)

FIGURE 2.6. During rainfall events functional landscapes infiltrate and store water, and if *reserves* of soil moisture are adequate, they respond with (a) *pulses* of growth in vegetation and (b) plant reproduction.

that is, the amount of moisture stored in the soil. If adequate, soil moisture initiates pulses of plant, animal, and microbial growth, for example, pulses of new plant leaves, stems, roots, and seeds, larger and more animals, and increased microbial activity. If the products from these growth and reproduction pulses stay within the landscape, as is largely the case in functional landscapes, they represent feedbacks and gains to the system as seen, for example, in a thick growth of perennial grasses such as spinifex (figure 2.8b).

All landscapes, even highly functional ones, also lose some resources. Earlier we noted losses of runoff from landscapes. Some products gained from growth and reproduction pulses are also lost from the landscape by what we call *offtake* processes. For example, fire is an offtake process that can dramatically reduce the amount of standing plant biomass (figure 2.8a). Two other major offtake processes that affect landscapes around the globe include the removal of vegetation by livestock grazing (figure 2.9) and tree clearing for agriculture (figure 2.10). Other offtake processes include, for example, game hunting and firewood harvesting.

(a)   (b)

(c)   (d)

FIGURE 2.7. Functional Australian landscapes provide habitats for thriving populations of wildlife: (a) kangaroos, (b) echidna, (c) wombat, (d) Tasmanian devils. Photographs courtesy of Timothy Blanche.

Feedback Processes: Building the System

Resource gains to a landscape initiate a number of biological and physical feedback processes (figure 2.11). Such processes are called *positive feedbacks* because they positively, or beneficially, affect components and processes within the landscape system. For example, a positive *biological* feedback process occurs when gains in plant biomass build the structural diversity of vegetation in the landscape (plate 2.3), which provides more habitats for a greater diversity of animals.

Biological feedback processes also enhance *physical* feedback processes. This is because an increase in the number, density, and diversity of vegetation patches covering the landscape surface (figure 2.11a) also improves the capacity of that landscape to retain more rainfall in the next storm event. Vegetation patches function by obstructing more flows and allowing more time for more water to soak into the soil, which can ultimately

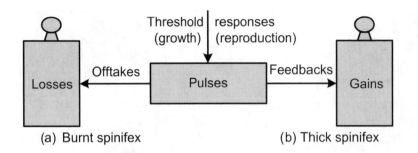

(a) Burnt spinifex   (b) Thick spinifex

FIGURE 2.8. In a landscape system, *pulses* of growth and reproduction represent *losses* if, for example, (a) spinifex grassland biomass is removed in a wildfire, but *gains* if, for example, (b) there is an increase in the thickness of spinifex.

(a)

(b)

FIGURE 2.9. Landscape offtake occurs when cattle consume forage and are then taken off the landscape and sold.

FIGURE 2.10. Landscape offtake occurs when rainforest is cut and burned to grow crops that are sold out of the area.

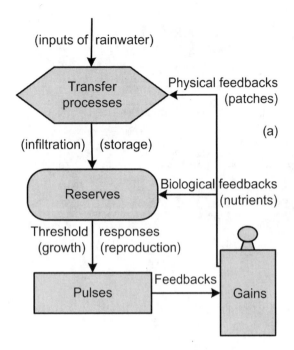

(a)

FIGURE 2.11. Pulses of growth and reproduction represent feedbacks or gains to the landscape by building, for example, a higher density and diversity of vegetation patches, (a) which obstruct flows of water during future rainfall events (a physical feedback process), and also store larger reserves of soil nutrients including organic carbon (a biological feedback process).

stimulate more growth of plants, animals, and microorganisms.

A *Fluctuating but Balanced Dynamic System*

Landscapes are continually changing as inputs that drive the dynamic processes that influence *gains* and *losses* in these systems fluctuate over time and space. We have depicted this by having gains and losses being balanced on a fulcrum (figure 2.12). Obviously at various times of the year, such as during a dry spell, landscape systems lose more than they gain, but this tilt in the balance reverses during rainy periods. In the long term we can view a landscape system as balancing gains and losses, while fully acknowledging that these gains and losses fluctuate over the short term as, for example, between seasons. In an earlier book (Ludwig et al. 1997), we illustrated how gains and losses become out of balance in dysfunctional landscapes.

Further Thoughts

In ending this chapter we feel that it is important to note some additional points about our framework. Our first point relates to the axiom mentioned earlier: "You can't manage what you can't measure." We used arrows between boxes to specify the dynamic processes that need to be measured in the landscape system. These processes must be measured in terms of changes in amounts and rates, for example, the rate at which rainwater infiltrates or soaks into the soil versus the amount that runs off the landscape. Processes are directly measured or, if difficult to measure, estimated by observing simple surrogates (indicators). The kinds of landscape processes that are directly measured versus those that are estimated by indicators will become evident by the examples we present in later chapters where we describe what monitoring data need to be collected to evaluate trends in restoration.

Second, we feel that a focus on processes enables RPs to apply field assessment or monitoring proce-

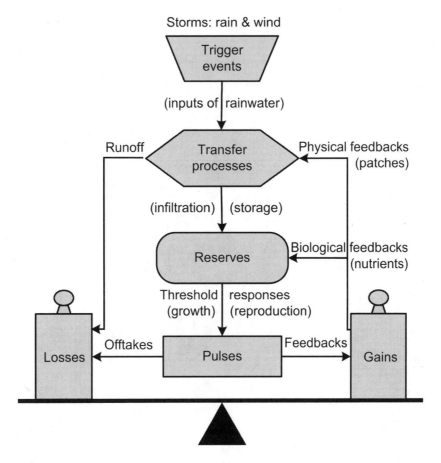

FIGURE 2.12. A completed conceptual framework depicting how functional landscapes, when triggered by events such as rainfall, respond in space and time with processes that transfer water by runoff and storage in soil reserves, which then initiate pulses of growth that are gained or lost by the system. The landscape system sits on a fulcrum to represent the fact that, in the long term, internal gains and external losses are dynamically balanced.

dures to a wide variety of landscape types, from rainforests to deserts. Although each landscape type has its own spatial scale and functional identity, they can all be described in terms of our conceptual framework; this is because they all have similar general components (plants, animals, microorganisms) and processes (e.g., infiltration, storage, growth pulses). The conceptual framework enables the information contained in monitoring data to be immediately appreciated by RPs, which helps them to read the landscape.

Finally, we feel confident that our landscape function-based approach, with its focus on processes, will lead to an understanding of how landscapes can be restored to be sustainable in the long term. We feel that there has been too much emphasis on restoring species rather than repairing fundamental landscape processes, which is putting the cart before the horse. Obviously the species present in a landscape are important, but species are most usefully expressed in terms of their role in landscape processes and in providing a diversity of functions.

Chapter 3

Principles for Restoring Landscape Functionality

Principles are important demonstrable assumptions or laws about the way a system works. Principles have been widely developed for systems that affect our daily lives, such as how to process and market healthy foods. In science, numerous principles have emerged from various studies of physical, chemical, and biological systems, such as how molecules move and how organisms evolve. From our studies of landscape rehabilitation, a number of principles have emerged that, when put into practice, help restoration practitioners (RPs) achieve their goals.

In this chapter we describe four of these principles, which we have tested in a wide range of environments from arid lands to tropical rainforests. They are an extension and a practical outcome of our function-based framework and our five-step procedure for adaptively restoring landscapes, which we described in the two previous chapters. After describing these four principles in this chapter we will illustrate how they apply to examples of landscape restoration in chapters within parts 2 and 3. If these four principles initially appear too complex, we ask readers to recall the dictum "For every complex question there is a simple answer: neat, plausible, and wrong" (H. L. Mencken).

As we noted earlier, our examples will usually emphasize processes involving water dynamics. This is not just whimsy; we choose water for our examples because it is a crucial driver in all living systems, whether it comes in the form of a gentle rain or a raging flood. Furthermore, based on our experiences, and those of many RPs, the failure of

damaged landscapes to effectively retain water for the benefit of plants, animals, and microorganisms is, universally, a key process that almost always requires restoration.

Principle 1: Analyze the causes of landscape dysfunction.

Our first principle essentially means to know your landscape setting and the underlying causes of your problem. To apply this principle, we ask questions such as what other landscapes are like in the area. Who has an interest in the specific area? What are the restoration goals? What, exactly, disturbed the landscape causing problems that initiated the need for a restoration project?

Why some landscapes have become damaged is very obvious, considering, for example, impacts caused by disturbances due to mining operations (plate 3.1). However, many problems have a range of causes that began many decades or centuries ago and have slowly accumulated over the years, such as those caused by long-term overgrazing in rangelands.

We emphasize that it is critical to fully understand how disturbances affect landscape processes, for example, how well landscapes retain water. We know that undisturbed landscapes strongly regulate overland flows of water because they typically have a high density and cover of perennial shrubs, grasses, and forbs (plate 3.2). In contrast, we have observed how a nearby damaged landscape with almost no vegetation cover (plate 3.3) is ineffective in regulating runoff within, and out of, this landscape. We also know that analyzing the causes of landscape dysfunction will reveal those properties and processes that are beneficial to the system, which helps RPs design and select the most effective restoration technologies and actions.

To analyze the causes of landscape dysfunction, we need to consider all of the components and flows (boxes and arrows) in the conceptual framework described in chapter 2. (See figure 2.12.) This enables us to systematically identify which processes have become ineffective within the landscape system and which have retained significant function. Have some processes simply ceased or have they been reduced in effectiveness? For example, in plate 3.3, it is obvious that biophysical feedback processes regulated by the plants have simply ceased. This leads to the next question: What physical and biological technologies can be applied to restore the processes involved in retaining water within damaged landscapes?

Principle 2: Restore ineffective processes sequentially.

Our second principle, which we have found to be essential, basically says that to restore damaged processes, such as the capacity of a landscape to retain water, apply physical and biological technologies in an ordered sequence. This means that you first apply technologies to improve those physical processes that function to retain water in the landscape. For example, compacted soil surfaces can be deep-ripped by machinery to loosen surfaces so that rainwater readily soaks (infiltrates) into the soil. Then you apply technologies that improve biological processes that enhance water retention in the landscape. For example, sowing seeds or planting seedlings on loose soil surfaces establishes a vegetation cover that functions to reduce raindrop impact, to slow flows of runoff, and to promote infiltration; all these help to retain water and soil in the landscape.

We have found that this principle of first restoring physical processes followed by biological

processes applies universally to all cases of successful rehabilitation of seriously disturbed landscapes. We will document this principle repeatedly by presenting graphs of data for indicators of rehabilitation progress in later chapters. Recall from chapter 2 that we defined indicators as simple surrogates for difficult-to-measure landscape processes. Data trends for biological indicators reflecting the development of landscape functionality typically follow an upward or increasing S-shaped (sigmoidal) curve with time (dashed line; figure 3.1). Indicators of physical processes typically follow a downward or declining S-shaped curve (dash-dot line). Net or whole landscape development is an upward S-shaped curve (solid line). Although rates (slopes of the curves in figure 3.1) vary, the contribution of physical technologies to improve functions such as water retention diminish as biological processes come increasingly into play and largely take over as the landscape progresses toward becoming a fully functional, self-sustaining system (leveling off of curves after time).

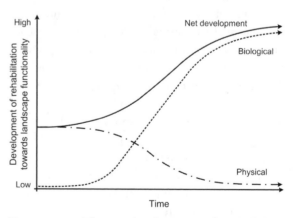

FIGURE 3.1. The net development of a rehabilitated landscape toward becoming a highly functional, self-sustaining, dynamically balanced system initially depends on applying technologies to improve physical processes (e.g., water retention), but over time the role of these processes is largely taken over by self-replacing biological processes (e.g., plant growth and reproduction).

The importance of applying principle 2 has also been confirmed from our studies of cases of unsuccessful landscape rehabilitation. For example, we have observed that spreading grass seed or planting seedling trees on a rehabilitation site before applying physical technologies to ensure that the site has the capacity to retain an adequate water supply for establishing and growing these plants is bound to fail. In other words, it does not work to put the cart before the horse.

Although examples of the application of principle 2 will be presented in subsequent chapters, we feel this principle is so important that the sequence of biological processes taking over from physical processes needs to be briefly illustrated here. We use an example of rehabilitation of a landscape damaged by mining. Although hypothetical, this typical example is based on our extensive experiences with restoring mined landscapes (e.g., Tongway et al. 1997; Ludwig et al. 2003; Tongway and Ludwig 2006).

For step 1 in our procedure (see chapter 1, figure 1.1), we simply assume that the goal is to restore the disturbed landscape so that it does not lose excessive amounts of runoff during and after rainstorm events. After applying steps 2 and 3 (problem analysis and designing solutions), two physical technologies were selected and applied (step 4) to enhance water retention in the landscape. First, the mined landscape was physically reshaped into a new landform using a design that minimizes runoff rate and sediment transport, which is to reduce slopes to a minimum (figure 3.2). Second, these slopes were deep-ripped along contours to reduce the compaction caused by the heavy machinery used in land forming. As seen in figure 3.2, deep-ripping forms banks and troughs, which are known to be very effective for improving landscape processes such as increasing rates of water infiltration and the amount of water stored in the soil. If it is necessary to form steep slopes, then to reduce runoff and erosion, large banks and troughs are constructed along slopes (figure 3.3).

At this point we note that there is an extensive literature on post-mining landform design and construction, which is beyond the scope of our book. We refer readers to a paper by Loch et al.

FIGURE 3.2. A landscape where piles of waste rock were reshaped to have minimum slopes. The reshaped surface was then deep-ripped to create banks and troughs.

FIGURE 3.3. A landscape reformed from piles of waste rock into one with sloping sides and a series of banks and troughs shaped along slope contours.

(2006) because of their successful and award-winning approach to landform rehabilitation. Their "key performance indicators" approach lists five components of successful landform rehabilitation that are very similar to, and confirm, the usefulness of our adaptive landscape restoration procedure.

After the mined landscape was treated by applying physical reforming and ripping technologies, biological technologies were applied, in this case, by establishing vegetation in troughs (figure 3.4). This is where water from light rain showers tends to accumulate. Establishing vegetation in troughs was achieved by first applying fresh topsoil stored during mining operations. (See chapter 4.) Using fresh topsoil with viable seeds helps ensure that the final species composition of the vegetation is similar to that of nearby native landscapes. To improve plant growth, fertilizer was applied because soil analyses indicated nutrient deficiencies. As plants grow, they produce litter that accumulates within troughs and decomposes to develop soils with greater infiltration rates.

After these physical and biological technologies were applied in sequence to rehabilitate this hypothetical landscape, key indicators were monitored and trends were evaluated. (See figure 1.1, step 5.) Trends indicated that water was now being retained in the landscape. In actual rather than hypothetical cases, such findings would verify that restoration goals were being achieved.

Principle 3: Monitor indicators reflecting
    landscape processes.

As noted at the end of the previous example, monitoring is required to evaluate how well a rehabilitated landscape is progressing toward desired goals. Monitoring addresses a number of important questions, such as what evidence confirms that the technologies selected and applied are being effective in restoring the damaged processes? What can be measured to provide this evidence? Is progress occurring at an appropriate rate? In answering these kinds of questions over the years, we formed principle 3, which emphasizes the critical importance of mon-

FIGURE 3.4. Vegetation has successfully established in troughs on a slope.

itoring or measuring simple, yet quantitative, indicators of processes over time that provide data on how well the landscape is functioning.

Although some attributes of landscape functionality are easily and directly measured, such as the number and size of erosion gullies found along a slope, other landscape attributes and processes are very difficult to directly measure. For example, measuring actual rates of water infiltration into soil following rainstorm events is not only difficult to directly measure but, even if easily measurable, would only provide a brief snapshot of a landscape process. As implied by principle 3, we strongly recommend measuring simple indicators reflecting landscape processes that can be readily monitored over time. We have found that principle 3 greatly helps RPs with designing restoration technologies and with monitoring and evaluating the outcomes of putting these technologies into practice.

There are a number of simple indicators that reflect landscape processes and that can be easily measured using simple methods and inexpensive tools. For example, the number and size (length and width) of perennial vegetation clumps (patches) can be readily measured along transects (defined by a measuring tape) on a site (figure 3.5). Vegetation patch density and size indicate the potential capacity of the landscape to retain water after storm events. The functional role of different plant species can be indicated by measuring the shape and thickness of the foliage comprising different types of vegetation patches, which also function as habitat for a variety of fauna. We place far less weight on indicators such as the absence of species, which may simply show that a problem exists. The kinds of indicators we and others have found most useful for monitoring landscape functionality will become evident as we describe examples in later chapters.

The potential for soil surfaces on the landscape to infiltrate water can also be estimated by a small set of simple indicators derived as part of a monitoring procedure known as LFA, short for *landscape function analysis*. The LFA procedure includes measuring surface erosion features and many other indicators of how well landscapes are functioning. (See chapters in part 4.)

FIGURE 3.5. Measuring the size of vegetation clumps along a transect.

Principle 4: View landscape functionality as a continuum.

We have found from experience that another useful restoration principle is to view landscape functionality as a continuum. This view is important because an analysis of landscape function/dysfunction involves dynamic processes and fluctuations that contribute to the balance of gains and losses from a landscape system. (See figure 2.12 in chapter 2.) Although all landscape processes are linked within the system, they all respond to disturbances in different ways and at different rates; some may respond quickly and strongly to disturbances while others may change only slightly. The net system response results in a *sliding scale* effect, because each process responds in its own way with increasing disturbance, thus producing a continuum from very dysfunctional to highly functional landscapes. Because our

adaptive landscape restoration procedure measures a diverse range of indicators of landscape processes and their responses, assessing these responses along a continuum of landscape functionality has proven to be a very useful way to view the progress of landscape restoration.

Another way to view this continuum principle is to think of step 5 in the adaptive landscape restoration procedure. (See figure 1.1 in chapter 1.) Step 5 is where indicators reflecting landscape processes are monitored over time until sufficient data are available to robustly evaluate trends in these data. These trends are then assessed as to whether they are moving along a continuum toward values expected for a highly functional landscape.

A useful graphical approach is to show photographs illustrating changes in landscape functionality along a continuum (figure 3.6a). These photographs

(a) Photographs illustrating vegetation development along a continuum

1973   1989   2002

dysfunctional   (Continuum)   functional

(b) Time-marks along a continuum of landscape functionality

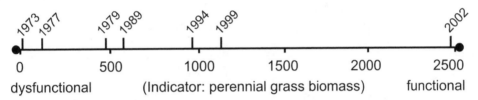

FIGURE 3.6. Photographs taken at the same photo point illustrating vegetation development from a very dysfunctional landscape in 1973 (top left), to intermediate functionality in 1989 (top center), and to a highly functional landscape in 2002 (top right). Cattle grazing and fire were strictly controlled to restore this savanna landscape. Time-marks for a perennial grass indicator (biomass as kilograms per hectare) were measured to reflect landscape restoration over time, which aimed to promote the retention of water and to reduce erosion by water and wind.

validate the axiom that a picture is worth a thousand words. It is also useful to plot trends in landscape functionality as *time-marks* along a line representing a continuum (figure 3.6b). Marks (vertical bars) noting the time (year) that indicators were measured are positioned along a continuum; these time-marks document any trends in the monitoring data.

Further Thoughts

We finish chapter 3, which is the last chapter in part 1, with a few additional points. The first is probably obvious but worth noting. Restoration technologies need to be applied at a scale appropriate to the extent of the problem. We have observed that in some cases treatments have been applied to only a small landscape area when the damage is quite extensive. Although less common, we have also seen treatments applied to extensive areas when the problem was actually localized.

Related to this point is the need to identify where to apply restoration treatments. In many cases we know that the cause of damage to processes is located upslope from where the signs of damage are actually observed. For example, a gully on a lower slope is actually caused by excessive runoff from upslope. By applying restoration technologies to these upslope areas, RPs can more efficiently and less expensively treat the problem.

Tackling landscape restoration problems with a multidisciplinary team is very useful. The ideal team has researchers and RPs working together, because restoring landscapes requires information from a number of sources. For example, understanding the drivers and processes involved in how landscapes function to retain water requires contributions from people in a wide range of disciplines, from geology, geomorphology, hydrology, and ecology to microbiology. A thorough analysis of different kinds of landscape data can help RPs select and apply those technologies that are most likely to achieve restoration goals.

PART II

Case Studies on Restoring Landscapes: Mine Sites and Rangelands

Landscape restoration presents an apparently bewildering variety of problems, especially when expressed in terms of renewing damaged ecosystem processes and restoring ecosystem components such as species biodiversity. To negotiate all these problems at specific sites, we want to provide restoration practitioners with a general approach, or template, that has a solid conceptual framework and rigorous principles underlying it. In part 1, chapter 1, we went through our five-step approach to adaptive landscape restoration. In chapter 2 we described a function-based conceptual framework forming the foundation for this approach, and in chapter 3 we defined four landscape restoration principles.

In part 2, chapters 4 and 5, our aim is to illustrate for restoration practitioners how to put our landscape restoration approach into practice by applying the conceptual framework and the four principles. We present four case studies where an adaptive landscape restoration procedure and the principles underlying this procedure were followed. We were directly or indirectly involved in projects described by these four case studies; our five-step adaptive procedure for restoring landscapes developed from these experiences. We selected these four case studies from a number of projects because they demonstrate most clearly the application of the adaptive landscape restoration procedure outlined in this book.

Three of these case studies are based in Australia and one in Indonesia. Our experience on other continents confirms that our landscape function-based approach is effective outside Australia and Indonesia as well. Restoring some landscapes is very difficult, of course, anywhere in the world. For example, we all know that repairing rangelands in arid or desert climates, especially where the landscape has deteriorated to shifting sand dunes, is extremely difficult and usually uneconomic. Restoring mine sites having severe geochemistry problems (e.g., acid drainage) can present restoration practitioners with nearly intractable problems.

In the introduction to each of the four case studies, we will describe the characteristics of the damaged landscape and the goals of the restoration process. We will discuss how other landscapes with similar characteristics are likely to respond positively to similar restoration efforts. These four case studies have progressed a long way toward achieving their restoration goals, so that we would say "trends are OK." (See figure 1.1 in chapter 1.)

Chapter 4

Restoring Mined Landscapes

We describe two case studies in this chapter on restoring mined lands. We first provide restoration practitioners (RPs) an example on how our five-step adaptive restoration procedure is being put into practice on landscapes disturbed by open-cut, bauxite (aluminum) mining operations. We present major findings from restoration projects we participated in on a mine site located on the Gove Peninsula in northern Australia; detailed results are provided in a report by Spain et al. (2009) and in a thesis by Wedd (2002).

Although this first case study is specific to Gove, we are confident that our five-step procedure also applies to restoring other landscapes where open-cut, surface-mining operations are being used to extract valuable resources, such as the examples we describe in chapters 6 (precious metals) and 8 (coal). Other examples of open-cut surface operations include mineral sand mining and pisolitic iron ore mining, although we do not describe these specific cases in our book.

In our second case study, we briefly describe for RPs how a rainforest landscape can be restored after mining for gold in East Kalimantan, Indonesia. One of us (DT) participated in monitoring this mine-site restoration project, which is described in detail in a thesis by Setyawan (2004). Although restoring mined landscapes in a high rainfall environment has some advantages over mine-site restoration projects in drier environments (see chapters 6 and 8), rainy landscapes pose other challenges, which we describe in this second case study.

Bauxite Mining, Gove Peninsula, Northern Australia

Aluminum is mined from geological deposits rich in hydrates of aluminum (bauxites). Thiry and Simon-Coincon (1999) estimate that 88 percent of the globe's bauxite is within lateritic deposits. Bauxite and laterite deposits formed during tropical and subtropical geological periods by surface and subsurface weathering processes. Laterite and bauxite are similar, but bauxite contains more aluminum and iron. Where bauxite deposits occur near the surface, they are readily mined. These deposits mostly occur in equatorial and subequatorial regions in South America, Africa, India, Southeast Asia, and in northern Australia (e.g., Gove Peninsula and Cape York).

Operations to mine lateritic bauxite are very similar around the globe. First vegetation (typically savannas, woodlands, and forests) is cleared away. Topsoil is then stripped off and moved to an area that is ready for rehabilitation. Then the layer of bauxite is excavated, dumped into trucks, and transported to a processing plant. These plants are usually located near a port so that either the bauxite or concentrated products such as alumina can be shipped to refineries to extract the metal; these refineries are located in places where electricity is less expensive, such as Iceland.

Study Area

The mined landscape being restored is located near Nuhlunbuy on the Gove Peninsula in northern Australia (figure 4.1). The region is seasonally humid. About 70 percent of the average annual rainfall of 1,444 mm (Gove Airport) occurs in the December to April wet season. The dry season from May to November is such that the average annual pan evaporation of 2,153 mm exceeds annual rainfall. When it is wet, it is very wet, and when dry, very dry. Temperatures on the Gove Peninsula are mild throughout the year, largely because of the influence of winds off the Gulf of Carpentaria to the east. In December the wet season maximum and minimum temperatures average 33°C and 25°C, respectively. In August the dry season maximums and minimums average 28°C and 19°C, respectively.

The natural vegetation is grassy open forest or savanna (figure 4.2), which is dominated by two eucalypt tree species, *Eucalyptus miniata* and *E. tetrodonta*. This vegetation type is common across northern Australia on well-drained sites with deep, red lateritic sands and gravels. Although the understory is typically grassy (tall tropical grasses), some palms and shrubs occur in the mid- to lower-canopy layers.

Soils are highly weathered laterites, rich in oxides of aluminum and iron. The lateritic topsoil is largely composed of large soil aggregates (plate 4.1) and numerous small roots and silica-rich, pea-sized nodules called pisolites (figure 4.3). The soil surface readily takes in rainwater (i.e., high infiltration capacity). The Gove Peninsula has a layer of bauxite about three to five meters deep below the topsoil in a number of locations, and these layers are mined.

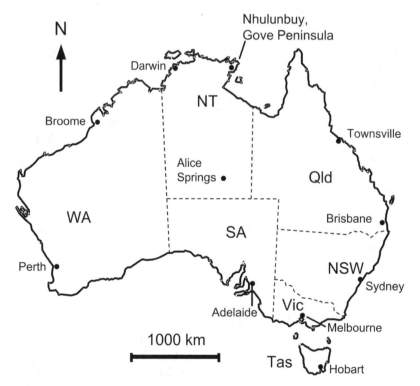

FIGURE 4.1. Location of the case study near Nhulunbuy on the Gove Peninsula in the Northern Territory, Australia.

FIGURE 4.2. A typical grassy open forest (savanna) in northern Australia, which serves as a reference area for evaluating rehabilitated sites. Photograph courtesy of Sue Gould.

Step 1: Setting goals

Landscapes being mined for bauxite at Gove are leased from the local traditional indigenous owners of the land. Restoration goals are specified in the lease agreement between the traditional owners and the mining company. The primary restoration goals in this agreement are to repair mined landscapes so that (1) landforms are stable and blend in with surrounding natural landscapes; (2) restored vegetation is self-sustaining and is similar to that of nearby natural forests; and (3) the traditional uses of these landscapes by indigenous people can be reestablished, including the preservation of totemic species and the establishment of specified traditional food plants.

Landscape restoration projects are an integral component of mining operations at Gove, and our role was to work with the ecologist-restorationist Mr. Dieter Hinz to specifically evaluate whether

FIGURE 4.3. A typical soil profile under grassy open forests on the Gove Peninsula showing the friable soil structure and the abundance of pisolites in the profile.

trends in vegetation and soil development on restored sites were toward those expected from natural sites (Spain et al. 2009).

Step 2: Defining the problem

At Gove, lateritic bauxite is mined on about 125 hectares each year. About the same amount of area has been rehabilitated each year since 1973 because the mining company has a policy of continuous restoration. There is a running five-year plan, so that mining and restoration can be harmonized over time. Unlike many mine sites, restoration procedures at Gove have been relatively consistent since mining started. This is because essentially the same system of rehabilitation has been applied by Mr. Hinz and his team since mining operations began in 1972. This system conforms to our five-step adaptive restoration process and its underlying principles.

Mining bauxite at Gove, as in other strip-mining operations around the globe, uses large machinery to push vegetation into rows, which are burned after drying (plate 4.2). This cleared landscape is not mined for about two years. During this time some vegetation grows from root-sprouts and seeds, which maintains soil biological activity. When mining starts, the topsoil (about 150 millimeters in depth) is stripped from the surface and is hauled to, and spread over, a new site being prepared for restoration. If no new sites are available, the stripped topsoil is hauled to a lay-down area. Mr. Hinz emphasizes the importance of "laying down" topsoil rather than stockpiling it. Laying down topsoil (spreading it in many small piles over an area) helps retain the biological vitality of the soil, whereas stockpiling typically creates a deep pile of soil, the interior of which dies from lack of oxygen and water.

The subsoil below the removed topsoil is then stripped and pushed into piles adjacent to where the bauxite layer (typically in 2 to 5 m thick layer below the subsoil) is then mined. Extracting the bauxite layer lowers the landscape and exposes a surface that is typically a hard, iron-rich laterite (plate 4.3), which is aptly referred to as ironstone. RPs face a difficult challenge when restoring such ironstone landscapes, but at Gove, Mr. Hinz and his team have a number of conditions working in their favor:

1. There are no steep slopes to deal with because landforms on the Gove Peninsula are subdued (i.e., not hilly or mountainous).

2. There are no highly toxic minerals to dispose of (e.g., sulfur-bearing rocks) because surface geological layers are relatively benign iron and aluminum oxides (i.e., laterites and bauxites).

3. When topsoil is stripped from surfaces, it can be managed to preserve its biological quality.

4. The Gove climate is favorable. As noted in the study area description, temperatures are moderate and there is a reliable wet season. These factors make it easier to regenerate self-sustaining vegetation because native plant seeds can be sown at the start of the wet season when they will quickly germinate and rapidly grow, and eventually produce more seeds.

Before designing a specific restoration project, Mr. Hinz first analyzed a problem that relates to principle 2: restoring the capacity of the landscape to retain water. Because natural landscapes in the area are very gently undulating, the mined area can readily be reshaped with machinery into a similar gently undulating landscape. However, as noted, after mining the landscape has been lowered by a few meters and an ironstone layer is exposed in most places. Mr. Hinz examined mined areas to see where ironstone layers needed to be deep-ripped to open up the surface and produce deep fractures. This deep-ripping allows water to infiltrate deeply into the soil, which promotes plant roots to grow deeper. To reduce compaction by machinery after subsoil and topsoil were moved back onto the ironstone surface, Mr. Hinz had this surface tilled (figure 4.4).

At this point we reemphasize that the function of resource retention, initially enhanced by physical processes such as surface ripping, needs to be eventually taken over by self-sustaining biological processes. (See figure 3.1 in chapter 3.) To promote biological processes, Mr. Hinz planted the site with native, perennial, self-replacing vegetation, which then functioned to retain and utilize resources within the landscape. To determine what

FIGURE 4.4. A landscape prepared for sowing (age zero) by reshaping, deep-ripping (after returning subsoil), and surface tilling (after returning topsoil). Photograph courtesy of Sue Gould.

kinds of plants to sow on mined landscapes, Mr. Hinz analyzed the plant composition of nearby natural open forests.

To protect surfaces from wind and water erosion, Mr. Hinz also examined natural landscapes and found that the function of surface protection was largely provided by a groundcover of native perennial plants, mostly tall tropical grasses. He also conducted experiments to see if the seeds of native plants remained viable in topsoil after being stored in the field for up to two years. If seed remained viable, a simple solution would be to spread fresh topsoil from a site being stripped, or older topsoil from lay-down areas, onto sites being prepared for rehabilitation. In addition, because local trees typically germinate under shade, Mr. Hinz created shady microclimates, using both local grasses and a hybrid annual sorghum. These microclimates facilitated a high germination rate of tree seeds and growth into saplings in the first year (plate 4.4).

Mr. Hinz also conducted soil analyses to identify any nutrient-deficiency problems. Northern Australia's tropical soils tend to have low concentrations of plant-available phosphorus due to having high concentrations of oxides of iron and aluminum; these oxides tightly hold phosphate ions so that they are unavailable for plant growth. If deficient, a small, single application of phosphate fertilizer could be applied, but this was considered a minor and inexpensive problem because the native plants being sown are well adapted to growing in low phosphorus soils.

A critical process on mine restoration sites is soil development (Spain et al. 2009). Soils are needed that are favorable for retaining rainfall and organic matter on-site and for storing and cycling of nutrients. Feedbacks from pulses of plant growth effectively promote soil development by biological and physical feedback processes. (See figure 2.11 in chapter 2.) With time, soil properties and processes should develop along a continuum toward becoming highly functional. (See figure 3.6.)

Mr. Hinz examined nearby natural forest sites and concluded that the species composition and the size of native trees found on these sites would be useful indicators of how well the vegetation was developing on restoration sites. Mine sites are often planted with *Acacia* species because they germinate readily and grow rapidly, and they notionally provide important ecosystem services, such as fixing soil nitrogen, adding litter to surfaces and organic matter to soils, and providing shade and shelter for a diverse biota. However, most of the acacias used for rehabilitation in northern Australia are sourced from areas outside the region. These acacias, which commonly occur as thickets in forest and savanna landscapes, are typically fire sensitive, that is, they are readily killed by the wildfires that occur in the dry season when grassy fuels are thick and highly flammable. But, native eucalypt trees in northern Australia are highly fire resistant when mature, and eucalypts form self-sustaining open forests, which are needed to meet restoration goals. Long-lived eucalypt forests also provide a wealth of goods and services that far outweigh those provided by short-lived acacia thickets.

Recall from earlier in this chapter that indigenous people on the Gove Peninsula greatly value certain tree and shrub species, and restoring these species is an important restoration goal. For example, Darwin stringybark (*Eucalyptus tetrodonta*) is a tree that needs to be established on restoration sites because it is of great spiritual, medicinal, and practical value to local traditional owners. If the plant species valued by traditional owners proved to have low viability in the available topsoil, Mr. Hinz collected seeds from these species and either sowed them onto sites or grew them in pots in a nursery until ready to plant as seedlings.

Steps 3 and 4: Designing solutions and applying technologies

To successfully restore mined landscapes at Gove, Mr. Hinz used an adaptive process to design the most effective treatments. This process was built on findings from a number of experiments conducted during the early days of rehabilitation. For example, at first Mr. Hinz simply placed topsoil onto reformed mined areas without sowing any seeds collected from local native forests. However, he found from early trials that some of the vegetation developed from this topsoil was dominated by undesirable fire-prone acacias. This vegetation is not stable because acacias are killed by lightning-initiated wildfires, which are difficult to control. At Gove, human-started fire damage to rehabilitated areas was controlled by a system of firebreaks.

In other early trials, Mr. Hinz planted exotic trees, but these too proved to be unsustainable because of fire sensitivity. He learned that sowing seeds collected from fire-resistant native eucalypt forests was the most useful practice. In the 1970s, he trialed different seed mixes, rates of sowing, and fertilizer applications to establish the most effective applications.

Currently, Mr. Hinz and his team have put into practice the following treatments to restore mined landscapes at Gove:

1. The ironstone subsoil layer is fractured by deep-ripping with heavy machinery.

2. The subsoil originally located between the topsoil and the bauxite is returned from nearby piles to the site being rehabilitated.

3. Topsoil from a nearby site being cleared for mining, or from a lay-down area, is spread on surfaces and tilled to reduce compaction (figure 4.4).

4. A seed mix of native trees and grasses is sown and tilled into the topsoil surface.

5. A single application of 100 kg per hectare of superphosphate fertilizer is spread over sown surfaces.

6. The scheduling of seed sowing and fertilization is set to occur just before the onset of the wet season.

7. Vegetation development is assessed by measuring the species composition (e.g., eucalypt versus acacia trees) and the size (e.g., height and basal area) of trees on rehabilitated sites, along with the

abundance (e.g., number per hectare) of other tree and shrub species important to the indigenous people of the region.

8. The role of soil fauna is observed, especially termite activity because of their role in transforming plant litter and improving soil structure (Spain et al. 2009).

9. Soil development is assessed by conducting some specific soil chemical and physical analyses (e.g., organic matter, nitrogen, phosphorus) and by observing a number of soil-surface indicators, which are specified in a set of landscape-monitoring methods known as *landscape function analysis* (LFA). (See chapters 13 to 16.)

10. Fire is actively excluded from all rehabilitated sites by building and managing fire breaks.

Step 5: Monitoring and assessing trends
To assess how rehabilitated mine sites at Gove were progressing toward specified goals, we worked with Mr. Hinz and his team to collect data on indicators of vegetation and soil development. Here we summarize a few key findings from this work, which we selected from the results reported in Spain et al. (2009). When we measured indicators on restoration sites, we used what is called a *space-for-time substi-*

tution approach, where sites at different locations of different rehabilitation ages were measured during the same project period (2001–2). This approach was necessary because indicators were not repeatedly measured (monitored) enough times at individual sites to adequately assess site trends through time. Ideally, data for indicators should be collected a number of times at every rehabilitated site so that sufficient data are obtained to have confidence in the generality of any trends found.

We will first illustrate the development of the vegetation on rehabilitated landscapes using photographs taken at a sequence of sites restored in different years. Then we will present data on indicators of vegetation and soil development collected from these sites.

Initially, sites rehabilitated just prior to the onset of the wet season are completely bare of vegetation (figure 4.4), but, in the wet season, seeds rapidly germinated and seedlings successfully established so that a three-year-old site had well-developed tree saplings and grasses (figure 4.5). On a thirteen-year-old site, trees were more fully developed (figure 4.6). Also evident in this photograph is that small tree and shrub species are dead or dying because of self-thinning. A site at a rehabilitation age of twenty-six

FIGURE 4.5. Three-year-old rehabilitation site at Gove illustrating how tree and grass species have established.

FIGURE 4.6. Thirteen-year-old rehabilitation at Gove showing how dense tree and shrub saplings are dying due to self-thinning processes.

years had very well-developed trees, and the ground was covered with grasses and tree litter (figure 4.7). The appearance of the vegetation in this photograph is quite typical for natural forest sites in the region. (See figure 4.2.) The twenty-six-year-old site was the oldest restored mine site we studied.

This trend in vegetation development was confirmed by color photographs taken at other sites over a sequence of time (years) since rehabilitation:

- Tree seedlings and grasses successfully established on a one-year-old site (plate 4.4).
- Tree saplings were abundant on a seven-year-old site (plate 4.5).
- Plant litter and grasses extensively covered an open forest floor on a twenty-year-old site (plate 4.6), which is similar to natural forest sites (figure 4.2).

Although the establishment of native trees is evident in figures 4.5 to 4.7 and in color plates 4.4 to 4.6, we noted earlier in this chapter that the species composition of the trees is also important. Recall that our desired goal is to have a dominance of native eucalypt trees, not fire-prone acacias. At Gove, our data on basal areas (a measure of tree size) of eucalypts and acacias at restoration sites of increasing age showed that trends were, in fact, toward the desired goal of eucalypt dominance (figure 4.8). Note that at the twenty-year-old site, eucalypt basal area (e.g., *Eucalyptus miniata* and *E. tetrodonta*) was approaching the mean basal areas measured at a number of native forest sites. This is a positive finding for tree composition and size. Also note that eucalypt basal area at the twenty-six-year-old site was well below that expected from the trend line. This is because the basal area at this site was dominated by native noneucalypt trees such as *Brachychiton diversifolius*, *Callitris intratropica*, and *Ficus opposita*. Although the reason for this twenty-six-year-old site being dominated by non-eucalypts is unknown, the total basal area of all trees on the site was 21.3 m² per hectare compared to a mean of 15.7 m² per hectare on nearby unmined sites, thus devel-

FIGURE 4.7. Twenty-six-year-old rehabilitation at Gove having large trees and a thick layer of tree and grass litter.

opment of all trees on our oldest rehabilitated site was actually above that expected.

We found another positive result in the data: the height of developing eucalypt trees was toward that expected for trees measured in natural forests (figure 4.9). We also found that the height of acacia trees was declining as taller and older acacia trees were dying off, which is part of their natural cycle. This declining trend in acacia tree heights was toward that expected for acacia height data we collected on natural unmined sites.

In the first few years, the grass layer rapidly developed on restored mine sites at Gove, which we measured as the mean size of grass clumps (figure 4.10). Recall from chapter 2 that clumps of grasses help prevent erosion by obstructing flows of water and wind over the landscape surface. But after about five years the size of grass clumps reached a peak on rehabilitated sites and then declined toward that expected on natural forest sites. This decline in grass clump size is related to factors such as increased levels of shading as the tree canopy thickened over time and because trees take more of the available soil water and nutrients.

The cover of tree litter on the surface of restored sites steadily increased with site age so that sites older than fifteen years had litter covers of greater

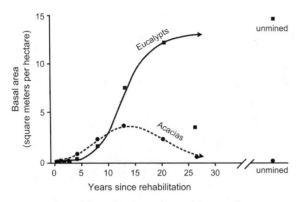

Figure 4.8. Basal areas of eucalypts and acacias on rehabilitated mine sites of increasing age and on an unmined site at Gove.

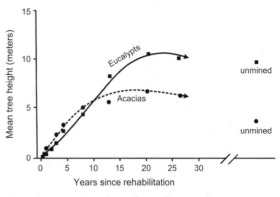

Figure 4.9. Heights of eucalypt and acacia trees on rehabilitated mine sites of increasing age and on an unmined reference site at Gove.

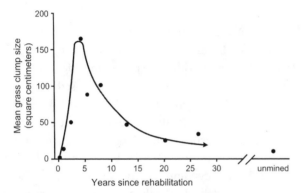

Figure 4.10. The mean size of grass clumps on rehabilitated mine sites of increasing age at Gove and on a nearby unmined site.

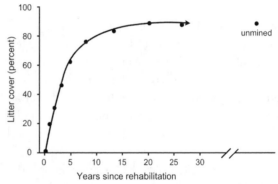

Figure 4.11. Mean litter cover on rehabilitation sites of increasing age and on an unmined site at Gove.

than 85 percent (figure 4.11) and showed abundant evidence of litter decomposition and incorporation into the mineral soil. These high values for litter cover were very similar to those for an unmined natural site that had not been burned. Sites recently burned by wildfires will have reduced covers of litter.

On rehabilitated sites we measured rates of water infiltration into soils using specialized equipment (Spain et al. 2009). We found that rates were initially about 1,000 mm per hour on freshly tilled soil but declined to about 200 mm per hour on a three-year-old site (figure 4.12). These low rates were related to the formation of surface physical crusts, and to the natural consolidation and compaction of the tilled surface. After three years, soils become much more porous to water, primarily because of the construction of *biopores* by soil macrofauna such as termites. Plants also produced litter to promote termite activity and other macrofauna, and plant canopies protected the soil surface from compaction by rainsplash. We found that mean infiltration rates on twenty- and twenty-six-year-old sites increased to about 3,400 mm per hour, which exceeded the mean value of 1,875 mm per hour we found on a nearby unmined site. This lower infiltration rate

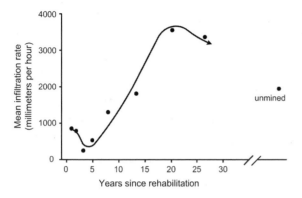

FIGURE 4.12. Measured water infiltration rates on rehabilitated sites of increasing age and on an unmined site at Gove.

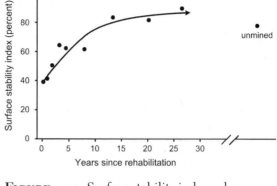

FIGURE 4.13. Surface stability index values assessed on rehabilitated sites of increasing age at Gove and on a nearby unmined site.

on a natural site is likely caused by frequent use of fire, which removes plant litter thus depriving termites and other soil macrofauna with the resources they need to provide ecosystem services, such as building biopores. Compared to water infiltration rates for soils in other regions of northern Australia, these rates are relatively high because of the permeable pisolitic soils found at Gove (figure 4.3).

Lateritic bauxite materials also form relatively stable surfaces. On rehabilitated sites at Gove we estimated surface stability using a synthetic index, which is derived from a number of readily observed indicators such as the amount of litter covering the soil surface. (See chapter 14.) The surface stability index is scaled as a percentage, where greater values indicate an increased resistance of the surface to erosion (i.e., stability). We found that a site of age zero had a mean surface stability index of about 40 percent (figure 4.13), which reflects the intrinsic resistance to erosion by mineral soil properties alone. This site had no vegetation (figure 4.4) and therefore had no biological feedbacks from plant growth pulses to improve soil-surface stability. But, as vegetation developed, we found that the surface stability index steadily increased so that a thirteen-year-old rehabilitation site had a stability index of about 80 percent, as did older sites, which exceeded that expected from an unmined site.

On each of the rehabilitated sites we studied at Gove, we also estimated a nutrient-cycling index, which is assessed with indicators of the effectiveness of nutrient-cycling processes and is expressed as a percentage. (See chapter 14.) For example, the nutrient-cycling index strongly draws on the amount and degree of incorporation of organic matter to form humus at the soil surface (e.g., the amount of litter and activity of decomposing organisms). We found that rehabilitated sites less than three years of age had nutrient-cycling indices of less than 20 percent (figure 4.14). The nutrient-cycling index then steadily increased to about 75 percent on a twenty-six-year-old site, which exceeded the expected value of about 50 percent on an unmined natural site. However, this lower value of 50 percent indicates that most of the natural open forests on the Gove Peninsula are frequently burned so that the litter layer is frequently lost, which results in a lower nutrient-cycling index and a reduction in the decomposition processes that improve soil biological quality.

We also directly measured some nutrient pool sizes in soil samples collected on Gove sites. We found, for example, that because superphosphate was added at the time of sowing, soil phosphorus on restored landscapes was higher (at all soil profile depths) than that found in nearby natural forests. (These data will not be presented here, but they are

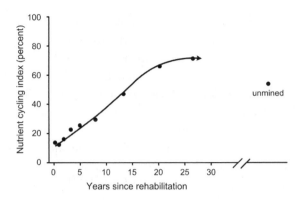

FIGURE 4.14. Nutrient-cycling index values for rehabilitated sites of increasing age and for an unmined site at Gove.

reported in Spain et al. 2009.) Rehabilitated sites showed a steady increase in nitrogen concentration and eight years of age had measured soil nitrogen contents (at the 0 to 1 cm depth) that exceeded those found in unmined sites, and after twenty years this trend held for soil samples collected down to 10 cm.

It is evident from photographs and an analysis of data from rehabilitated sites that, compared to unmined natural sites, restoration trends were OK. (See figure 1.1 in chapter 1.) This "trends OK" assessment applies for two of the three goals set to be achieved for mine-site rehabilitation at Gove. These two goals were to create stable landforms that blended in with surrounding natural landscapes and to restore self-sustaining ecosystems similar to those found in nearby natural forests. Our project did not specifically examine the third goal, which was to restore the landscape so that it would meet the material, cultural, and spiritual needs of the local indigenous people. Confirmation of meeting this third goal can be seen in the 1996 video *Walyamirri: Return of the Living Environment* (Dieter Hinz, pers. comm.). This video indicates that local Aboriginal art from about 1990 on incorporates symbols of a partnership between indigenous people and the miners now working on the Gove Peninsula.

Fire had been excluded from studied rehabilitated sites at Gove from the beginning of restoration, so questions remain as to how well restored mine sites will resist damage from wildfires and severe cyclones, and how quickly vegetation and fauna such as birds will recover from such disturbances. These questions are beginning to be addressed (e.g., Ross et al. 2004; Gould 2010).

Gold Mining, East Kalimantan, Indonesia

Although the discovery of gold nuggets along surfaces of stream beds ignited famous gold rushes such as in California (1848) and in Victoria, Australia (1851), today gold is primarily mined from subsurface ore bodies. Gold is also obtained as a by-product of mining for other metals such as silver, lead, zinc, nickel, and copper. Gold-bearing ore bodies are found around the world, notably in countries such as Australia, Brazil, Canada, South Africa, and the United States, and in other places where tectonic events have thrust these ore bodies close to, or onto, the earth's surface, especially along geologic faults. Except for recent tectonic events (e.g., volcanic eruptions), geologic intrusions of metal-rich minerals have undergone prolonged geological processes to form distinctive deposits.

Depending on the geologic process, three general types of gold deposits are formed:

1. Lode-gold deposits occur as veins of quartz (lodes, reefs) formed as intrusions within volcanic (e.g., basalt), sedimentary (e.g., turbite), or igneous (e.g., granite) rocks.

2. Laterite-gold deposits are formed when geochemical weathering processes modify gold-bearing rocks to create gold deposits within iron-rich laterites (e.g., in Western Australia).

3. Placer-gold deposits are formed by alluvial processes, such as along rivers, which create secondary deposits of gold (e.g., nuggets, flakes) by the

weathering and transport of primary deposits. Alluvial processes have also formed *deep-leads* of gold within ancient sedimentary rocks (e.g., Witwatersrand deposits in South Africa).

During historic gold rush periods, mining placer gold primarily involved panning, sluicing, and using jets of water to extract gold flakes and nuggets from alluvium deposits along rivers and creeks. These mining methods are still used in a few mines, but most large commercial gold mining operations use open-pit and subsurface (tunnel) techniques to reach rich, gold-bearing lode, laterite, and deep-lead ore bodies.

Although gold ore-body extraction techniques, and how ore is processed to obtain gold, vary in detail at different mines around the globe, a common feature is that they all disturb natural landscapes and create dumps of waste rock (i.e., piles of regolith, spoil, and low-grade ore) and tailings ponds (i.e., containment or storage facilities for slurries of the finely ground rock remaining after gold and other metals are processed).

Here, we briefly report on landscape-restoration findings after gold mining at Kelian in East Kalimantan, Indonesia. Mining at Kelian disturbs rainforest and creates piles of waste rock and tailings ponds, which are similar in general structure to those created at other gold mines around the world, but Kelian is tropical and mountainous. Because of our extensive work on mine sites in arid and semi-arid environments (e.g., Tongway and Ludwig 2007), our aim here is to demonstrate that our five-step adaptive landscape restoration procedure also applies to the wet tropics. For brevity we will present only a few findings from the extensive results in a PhD dissertation by Setyawan (2004), which documents landscape restoration trends at Kelian.

Study Area

Kelian gold mine is located near the equator in East Kalimantan, Indonesia. In 2004 it was the largest gold mine operating in Indonesia (Setyawan 2004), but it was heading toward closure. In the mountainous region of East Kalimantan the climate is wet tropical. The annual rainfall exceeds 4,000 mm and the wet season is almost continuous. The natural vegetation surrounding the mine site is thick rainforest (figure 4.15). Slopes typically exceed 30 percent in the mined area. Soil profiles are characterized by a dark topsoil (due to humified organic matter) and a lighter subsoil (due to leaching).

Step 1: Setting goals

At the Kelian site, the mine operators' primary goals were to prevent excessive erosion (i.e., to retain water and soils on-site) and to return landscapes disturbed by mining back to tropical rainforest so that local people could use these landscapes as they did before mining.

FIGURE 4.15. Natural rainforest near a mine site in East Kalimantan, Indonesia.

Step 2: Defining the problem

Restoring disturbed landscapes in wet, tropical, mountainous environments would seem to pose monumental challenges because high rainfall intensities can readily erode steep slopes exposed by mining. However, RPs can successfully restore rainforests in these landscapes provided problems are carefully analyzed relative to landscape-function concepts and principles (chapters 2 and 3). In these wet and steep landscapes, a team of RPs worked on the problem of retaining water and soil resources in disturbed landscapes. This essentially means that they needed to protect slopes disturbed by mining from erosion. To solve this problem, they analyzed and designed ways to apply principle 2, that is, how to sequentially restore landscape processes that have become ineffective.

Fortunately, the team found that the natural geological and soil materials on the Kelian mine site were highly porous to water intake, which helps physical infiltration processes. They found that topsoil had a low potential for physical crust formation and contained high concentrations of organic carbon. The team explored ways to save and manage this high-quality topsoil so that it would be available for restoring disturbed areas.

Steps 3 and 4: Designing solutions and applying technologies

When conditions became favorable (i.e., a break in rainfall), the team added biologically active topsoil to slopes disturbed by mining operations. They dumped topsoil in discrete heaps so that the bases of the heaps overlapped across the slope (figure 4.16) to form a checkerboard of *microcatchments*, each about 25 m^2 in area. This procedure has various names, including *paddock dumping* and *pimple dumping*, the latter name referring to the appearance

FIGURE 4.16. Topsoil has been dumped in discrete heaps to form a patchwork of microcatchments across a slope on a mine site in East Kalimantan.

of the landscape when dumping topsoil has ended. The team then planted rainforest vegetation, including a leguminous, rapidly spreading vine; this planting supplements the germination of viable seeds already in the topsoil.

Step 5: Monitoring and assessing trends
Mine sites at Kelian rapidly rehabilitated when the RPs put principles into practice to restore key landscape processes. They sequentially applied technologies to retaining water and soil resources on disturbed slopes by first physically creating microcatchments and then planting rainforest vegetation. During rains, the microcatchments held and slowed surface runoff so that runoff did not cause erosion rills and gullies down the slopes (figure 4.17). In just weeks, vines were growing on rehabilitated slopes (figure 4.18), further protecting these slopes from

erosion. After about twelve months of vegetation growth, the team (with help from local people) was able to physically clear small spots within the emerging tropical forest floor to plant seedlings of selected food species grown at the mine site's plant nursery. After only fourteen months, rehabilitated slopes were growing a diverse mix of rainforest plants (figure 4.19).

This photographic record indicates that the principles and technologies put into practice at the Kelian mine site have led to success in retaining rainwater and restoring rainforest vegetation. To confirm this apparent success, Mr. Setyawan sampled soils and measured vegetation on rehabilitated sites aged three months, one year, and seven years (Setyawan 2004). As noted early in this chapter, one of us (DT) worked with Mr. Setyawan to measure three indicators of soil-sur-

FIGURE 4.17. Restoration practitioners measuring surface condition indicators on "paddock-dumped" locations on a mine site in East Kalimantan (photo looking downslope).

FIGURE 4.18. Vines growing on a rehabilitated slope at a East Kalimantan mine site. A paddock-dumped pile of soil is clearly visible in the center of the photograph.

FIGURE 4.19. A mix of rainforest plants growing on a rehabilitated mine site in East Kalimantan fourteen months after commencing restoration.

face condition: stability, infiltration potential, and nutrient-cycling capacity. (See chapter 14.) Here, we present these findings, which show that all three indicators rapidly increased from three months to seven years (figure 4.20). The seven-year-old rehabilitated site already had indicator

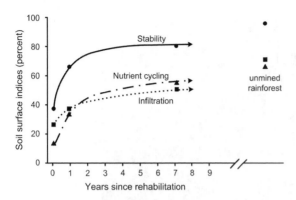

FIGURE 4.20. Trends in three soil-surface indicators on rehabilitated sites aged from three months to seven years, and in nearby rainforests, East Kalimantan.

values that were trending toward those expected for a nearby unmined rainforest.

Based on these findings for landscape-surface indicators, and other data described by Setyawan (2004), we could declare that landscape restoration trends were OK at the Kelian mine site after only seven years. But, this is premature because natural rainforests in East Kalimantan have trees up to 60 m in height, and other vegetation occurs in multiple lower layers. Although it will obviously take some time for trees on rehabilitated sites at Kelian to

reach 60 m and to form multiple layers, the primary goals were achieved: preventing excessive soil erosion and reestablishing rainforest vegetation. We would expect that in such a favorable environment, one could confidently declare successful rehabilitation in about twenty years.

Further Thoughts

In ending chapter 4, we want to emphasize our confidence in the global applicability of our five-step, adaptive landscape restoration procedure to mine sites. This confidence is affirmed by the two case studies described here, which were in different landscapes and climates, and because operations to mine lateritic bauxite and to mine gold are similar around the globe. When restoring mine sites, we have also found some "universal truths" that relate to our landscape function-based approach. For example, we have evaluated over fifty mine sites around the world and we found that those having the most restoration success were where RPs carefully managed their topsoil resource, no matter the climatic setting.

As you will see in the two case studies described in the next chapter, our restoration procedure is also effective in restoring disturbed rangelands.

Chapter 5

Restoring Damaged Rangelands

In this chapter we present two case studies on restoring disturbed rangelands. Our aim is to demonstrate how ranchers applied our five-step procedure, and its underlying principles, to repair arid and semiarid rangelands, respectively. Rangelands are nonarable areas used primarily, but not exclusively, for grazing livestock, such as cattle, goats, and sheep, and they cover about one-half of the earth's land surface.

Because of pressures from growing populations for more goods and services (food, water), use of rangelands is intensifying (see http://earthtrends.wri.org). Land-use intensification has caused extensive rangeland degradation, or what is commonly referred to as *desertification*. We refer restoration practitioners (RPs) to a book on global desertification (Reynolds and Stafford Smith 2002), because a discussion of the many biological, physical, social, and economic problems related to rangeland desertification is beyond the scope of our book.

Of all the rangeland degradation problems, loss of plant cover and soil erosion are two of the most extensive and severe types of damage observed around the world. In the specific case studies we present here, these two problems are tackled using an adaptive landscape restoration approach. Although our two case studies are in Australia, we are confident that this approach applies globally. This confidence is based on our many years of working on rangeland restoration (e.g., Tongway and Ludwig 1996, 2002a, 2007) and also on the work of our rangeland colleagues around

the world (e.g., Whisenant 1990; Milton et al. 2003; van den Berg and Kellner 2005; Kellner and Moussa 2009).

We first describe the details of a case in which a cattle producer on Woodgreen Station located in arid Central Australia has intuitively restored his arid rangelands (Purvis 1986). In this case, water-ponding technologies were successfully applied to repair rangelands damaged by sheet and gully erosion on slopes of up to about 2 percent. In our second rangelands case study, we briefly illustrate how water-ponding technologies were also put into practice to reclaim extensive areas of bare soil found in extremely flat, semiarid country (Thomson 2008). In both of these case studies, one of us (DT) collected field data as part of our role to assess restoration trends, which also relied on the work of rangeland colleagues (e.g., Bastin 1991).

Rangeland, Woodgreen Station, Central Australia

Europeans settled the arid lands of Central Australia in the late 1800s and early 1900s. They introduced domestic cattle to create commercial rangeland enterprises, or what are commonly called cattle ranches or stations. Initially these enterprises prospered as cattle utilized an abundance of natural forage consisting of palatable and nutritious grasses, forbs, and shrubs. Unfortunately, European rabbits were introduced into Australia in the late 1800s, and by the early 1900s rabbits had spread into Central Australia. As highly reproductive consumers of plants, rabbits, combined with droughts and over-grazing by cattle, removed the protective cover of plants across many landscapes, which greatly increased soil erosion.

Reduced profitability followed because of the lack of sufficient forage and water for cattle. After about 1910, running livestock profitably in Central Australia was only possible in good seasons. Stock often had to be walked many hundreds of kilometers to the nearest railhead (e.g., Alice Springs) to reach markets. Thus, in the first fifty years or so after settlement, stocking rates reflected what could be carried in good seasons (i.e., growing seasons with more than 50 mm of rain).

In the 1950s, to increase the supply of drinking water for stock, numerous bores were drilled into an artesian basin with huge water reserves across arid Central Australia (James et al. 1999). This meant that water was less limiting so that higher stock numbers were maintained during droughts. But now, stock constantly grazed and trampled pasture species, further reducing cover and accelerating soil erosion.

Once the extent of rangeland damage in Central Australia was recognized, attempts at repairing the landscape processes that had been damaged (e.g., retention of rainwater and soil) proved to be very unreliable, expensive, and slow. In hindsight, early restoration failures were largely due to RPs failing to fully understand how landscapes function. (See chapter 2.) In spite of these failures, there were some notable achievements. Ranchers with a keen eye for "reading" landscapes were successful in restoring damaged rangelands. Using their own resources, they thoughtfully analyzed their problems, trialed solutions, and adjusted technologies to achieve restoration goals (i.e., they intuitively put adaptive landscape restoration procedures and principles into practice).

Here we describe how a rancher, Mr. Bob Purvis, is successfully restoring his arid rangelands on Wood-green Station. Rangeland scientists worked with Mr. Purvis, including one of us (DT), to document his success. Our account of this case study will highlight key findings; details are provided in journal papers (Purvis 1986; Bastin 1991).

Study area

Woodgreen cattle station is located 180 km north of Alice Springs, Australia (figure 5.1). It was originally settled in 1915 and used to raise cattle and, for a time, large numbers of horses. Stocking rates were high and drainage lines were deliberately modified with earth diversion banks, which greatly altered runoff patterns. This had unintended consequences, such as the drying up of highly productive natural *floodout* (run-on) areas having the most fertile soils. For about fifty years, combinations of drought and grazing pressure destroyed pasture plants and initiated soil erosion on a massive scale (e.g., deep gullies). Mr. Purvis estimated that by 1965 about one million cubic meters of soil were lost, and gullies were continuing to cut upslope.

Step 1: Setting goals

To repair his damaged rangelands, Mr. Purvis set reclamation goals to reduce soil erosion by controlling runoff, return native perennial pastures in terms of both species composition and abundance, improve soil productivity, and maximize profits per animal. These specific goals were part of a five-point management plan he developed to restore Woodgreen as a productive pastoral enterprise (Purvis 1986):

1. To improve the productivity of damaged country
2. To establish a productive breeding herd
3. To adjust stock numbers based on the current condition of the better land
4. To set these stock numbers as if the year would be dry (i.e., stock conservatively)
5. To manage debt in the short term

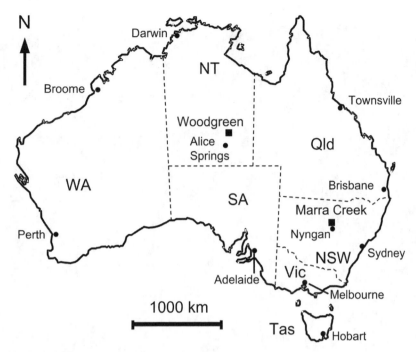

FIGURE 5.1. Location of Woodgreen, north of Alice Springs in the Northern Territory, and Marra Creek, north of Nyngan in New South Wales, Australia.

Although Mr. Purvis treated all five points as essential for his enterprise as a whole, here we will deal specifically with only his first point (improving damaged rangelands) because it provides a real-world example of our five-step adaptive landscape restoration procedure.

Step 2: Defining the problem

Mr. Purvis concluded that, historically, cattle numbers had been so high that disturbances due to grazing and trampling prevented any improvements in pasture production by natural restoration of landscape processes, such as healing of gullies. Rainstorms continued to cut gullies because runoff flowed far too rapidly down largely bare, smooth, and hard landscape surfaces. Reducing stock numbers was not the only solution, because Mr. Purvis saw that paddocks he had destocked for as long as twenty-five years had not naturally recovered.

Retaining water on the landscape became his top reclamation priority. But, his early attempts to cap-

ture water by using mechanical soil-surface treatments proved unsuccessful. For example, to collect water he tried tine-pitting where tractor-drawn machinery is used to dig small pits. Although such treatments were being widely applied around Alice Springs in the 1960s and 1970s, subsequent analyses found that such treatments generally fail to be effective in the long term (Friedel et al. 1996) because pits fill with sediment (figure 5.2).

To retain water and prevent further gully erosion, Mr. Purvis took a landscape-scale view. He concluded that large earthen banks to pond water (i.e., water-ponding technologies) were needed to control overland flows and to spread runoff across his most pastorally useful rangelands. The banks needed to be large enough to pond water upslope to promote infiltration and storage in the soil, and to survive extreme rainfall events without damage.

After rains Mr. Purvis observed that many Woodgreen soils would form a hard surface crust. One of us (DT) found that these crusted soils had water infil-

(a)

(b)

FIGURE 5.2. Landscapes with bare dispersive soils near Alice Springs where tine-pitting was applied as a restoration technology: (a) pits initially captured enough water to allow a few plants to grow, but (b) the shallow pits soon slumped and filled with sediment so that most plants died. Photographs courtesy Margaret Friedel.

tration rates of only a few millimeters per hour. During rainstorms only a small proportion of the rainwater was stored in the soil profile, and rates of runoff were high. Mr. Purvis concluded that mechanical treatments were needed to improve infiltration. He initially tried to open hard surfaces by deep-ripping dry soils with a single tine behind a bulldozer, but this failed to penetrate the hard soils and only lifted the back of a 25,000 kilogram bulldozer; deep-ripping had to be abandoned.

Rather than attempt to widely apply restoration treatments such as deep-ripping across damaged paddocks, Mr. Purvis focused his restoration efforts on smaller areas within paddocks where soil erosion was most active. He also selected areas where response to treatments would likely be most effective. He knew Woodgreen had a mixture of landscapes with different vegetation and soil types. He also knew from years of experience that some soils were inherently more fertile and productive than others. He recognized which of his more productive soils would most effectively respond to restoration technologies, such as water ponding; these were soils that swelled slightly, but significantly, on wetting, and then, on drying out, shrank to form cracks.

Soil scientists have shown that soils with small but significant swell/shrink properties have very positive responses to water-ponding technologies, because soil cracks form preferred infiltration pathways so that rainwater goes deep into the soil profile (Ringrose-Voase et al. 1989). Soils without swell/shrink properties are unlikely to respond positively to water ponding. Mr. Purvis noted that soils with crusted surfaces had very high water runoff rates and therefore did not wet adequately during rainstorms and the swell/shrink process failed. (The swell/shrink property of a soil can be assessed by using methods described in a paper on "A rapid method for estimating soil shrinkage" [McKenzie et al. 1994].)

Steps 3 and 4: Designing solutions and applying technologies

In designing technologies to control runoff and erosion to improve pasture production, Mr. Purvis

realized from his analyses that he needed to address two major problems: that the amount of runoff flowing into active gullies must be substantially reduced, and that runoff water must be held on areas of fertile and productive soil.

Mr. Purvis selected water ponding as the best solution because earthen banks could be designed and built to capture runoff and prevent it from flowing into gullies. Applying water-ponding technologies is a way of storing more water in the soil profile by extending the time available for water to infiltrate into soils, which is crucial when these soils have low infiltration rates. This technology also addressed the need to rehabilitate areas of severe damage such as gullies.

To pond water, Mr. Purvis initially constructed earthen banks by excavating or "borrowing" soil from across a slope to form a small dam about 200 m long. This technique created a ditch or borrow-pit upslope of the earthen bank. Although these dams captured runoff during rainstorms and provided water for animals to drink (positive outcomes), they also formed temporary deep ponds that water-logged soils and killed establishing pasture plants; new plants failed to grow on the exposed subsoil (negative outcomes).

Mr. Purvis changed his water-ponding design. He formed banks by pushing earth upslope from a borrow-pit below the bank (figure 5.3). Instead of exposing deeper subsoils in borrow-pits in front for the bank, which plants then had to colonize, the new design preserved the existing soils and any remnant vegetation upslope of the bank (plate 5.1). The bank was curved slightly to retain rainwater to a depth of about 150 mm, after which it was able to flow gently around the ends (spill points; figure 5.3). The banks were constructed to be about 2.5 m wide at the base and about 1.2 m high. The ends of banks were shaped like wings so that excess water would run gently down the hillslope well away from any gullies below the pond. The earth banks were created with a bulldozer, usually in the few days after significant rains when the soils were soft enough to work.

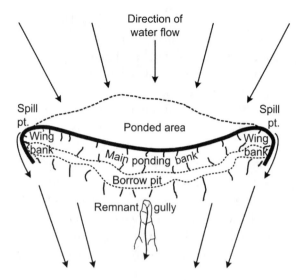

FIGURE 5.3. A sketch of the water-ponding bank design Mr. Purvis applied on Woodgreen.

Mr. Purvis learned that the size and shape of water-ponding banks is extremely important (figure 5.3). The curve of earthen banks was built sharper on steeper slopes to avoid banks being washed out during big storm events. Water-ponding banks must be sufficient in size to efficiently harvest water from large catchment areas, but these areas must not be too large because then big storm events would wash out earthen banks. Where water-ponding banks were positioned in the landscape was also very important. Mr. Purvis first built shorter, lower banks near the top of the watershed in paddocks. Then he built wider, higher banks farther down the watershed as the area of catchment increased. When feasible, Mr. Purvis built water-ponding banks on existing flats because he knew that soils on flats typically have swell/shrink properties and are inherently more fertile.

At first Mr. Purvis planted seeds of pasture species upslope of the bank where water was ponding (figure 5.4). Initially he sowed seeds of exotic grass species because the degraded landscape lacked local native perennial grasses. Seeds of U.S. cultivars of buffel grass, *Cenchrus ciliaris*, were sown because

buffel grass is known to be an effective colonizer and is palatable to cattle (Friedel et al. 2006). Later, Mr. Purvis stopped seeding water-ponding areas because native perennial grass seeds became sufficiently abundant as landscapes were rehabilitated.

Mr. Purvis knew that it is also important to manage total grazing pressure by cattle, horses, kangaroos, and rabbits so that pasture plants would establish on rehabilitated areas, especially within newly constructed water-ponding areas. On Woodgreen, rabbits are not a problem because soils are not suitable for what are called rabbit warrens; when rabbits try to dig burrows, soils either collapse or are too hard. Mr. Purvis first reduced the number of wild or feral horses. Then he reduced kangaroo numbers and adjusted his cattle numbers. Because all stock watering in pastures on Woodgreen is done with water troughs, water is supplied to cattle when they are in the paddock, but water is turned off when cattle are moved to other paddocks; this prevents kangaroos from using it. This water management technique effectively prevented high kangaroo densities.

Step 5: Monitoring and assessing trends

To evaluate the effectiveness of his water-ponding design, Mr. Purvis inspected banks for any damage after large rain events. If damaged, he immediately repaired and adjusted the size and shape of his earthen banks to prevent future damage. He also conducted a number of other monitoring activities, at times in collaboration with rangeland scientists.

To assess whether rehabilitated areas within pastures were developing a rich diversity of desirable forage plants, Mr. Purvis regularly monitored these areas by standing in place, rotating 360°, and counting the number of desirable pasture species. If he counted twelve to fifteen species in an area within a paddock with productive soils (used as country to fatten cattle), this indicated that the rehabilitated area was in good condition, but if he counted fewer than twelve species, it was assessed to be in poor condition. In less productive paddocks (used as country to breed-up cattle numbers), a count of eight to ten

FIGURE 5.4. An example of water ponding on hard soil above a recently constructed bank (off the photo to the right). The flooding followed a storm event of approximately 20 mm in November 1985. Photograph courtesy Gary Bastin.

desirable forage species indicated good condition and less than eight species, poor condition.

To evaluate whether ponded areas were establishing new pasture plants, Mr. Purvis examined the upslope boundaries of ponds to see if sediment was depositing and extending upslope, and if new plants were growing in these areas. He also examined the entire ponded area. If palatable perennial plants were well established, he used this information as part of his stock management strategy to make decisions about grazing the paddock. Mr. Purvis' stock management strategy is to rotate small herds of cattle through three paddocks by rounding them up twice a year so that each paddock is spelled for twelve months. He varied this pattern in dry times by moving cattle to other parts of his ranch where isolated

rainstorms have occurred and avoiding, if possible, moving cattle into a paddock that has had virtually no rain in the preceding twelve months. He aimed to keep cattle numbers low in paddocks being restored.

To assess whether gullies below water-ponding banks were "healing," Mr. Purvis looked for indicators such as whether sharp gully edges were rounding off, and if pasture plants were establishing within the gully. (See chapter 15.)

Finally, to evaluate whether soils were improving within ponded areas, Mr. Purvis worked with one of us (DT) to assess indicators of soil-surface condition such as friability and cracking (figure 5.5), which indicate a greater potential for higher infiltration rates after the soil surface dries and cracks. (See chapter 14.)

FIGURE 5.5. The soft, friable cracked soil surface that developed on a ponded area above a ponding bank constructed at Woodgreen.

Mr. Purvis evaluated these restoration indicators to assess whether water-ponding banks were effectively repairing his damaged rangelands. He also examined photographs taken at fixed photo points located at water-ponding banks and found positive trends. For example, pasture renewal at water-ponding bank number 6 is clearly illustrated by tonal photographs taken in 1985 and 2001 (figure 5.6) and by color photographs taken in 1985 and 1990 at bank number 5 (plates 5.1 and 5.2).

To confirm the improvements evident in photographs, Mr. Purvis, and one of our colleagues Gary Bastin (Bastin 1991) collected vegetation data from 0 to 80 m above four water-ponding banks of increasing age (0, 2, 5 and 15 years). They found that average perennial vegetation cover increased in an S-shaped trend (figure 5.7). However, more data are needed to reliably confirm this trend because average cover values were quite variable due to data being measured at one time period on four different aged water-ponding areas. In chapter 4 we defined this approach as *space-for-time* substitution. They also found that average grass cover was highest near banks, as might be expected, because this is where water most frequently ponds. Grass cover averaged 18 percent in the first 10 m from the bank, 7 percent from 11 to

40 m, and 0.4 percent from 41 to 80 m (at this distance water did not pond).

To determine whether soils were improving, they also sampled soils at the same four water-ponding banks by collecting soils at five places along five transects extending away from each bank. The first sample on each transect was taken from beside the bank wall and the others were collected upslope from the wall. They collected the final two soil samples beyond the area where ponding occurred. They analyzed soil samples for organic carbon, plant-available nitrogen, and plant-available phosphorus, using standard methods, which are described in Tongway et al. (2003).

They found that soil organic carbon greatly increased with age of the water-ponding bank in samples collected within 10 m of banks (figure 5.8). This increasing trend in soil carbon is likely to level off to an upper limit in the future, but this maximum level of carbon is as yet unknown. However, we can estimate the amount of organic carbon that was sequestered in soils near ponding banks over the fifteen years from 1985 to 2001. Assuming a specific gravity of 1,450 kilograms of soil per cubic meter on the fifteen-year-old site, the average amount of carbon (C) per square meter to a depth of 10 cm in soils sampled near the bank (within 10 m) was 697 grams of C per square meter compared to only 297 grams of C per square meter in soils upslope of the bank that never ponded. Over fifteen years, this represents an increase in stored carbon of 400 grams of C per square meter (or 4,000 kilograms of C per ha) in the top 10 cm of soils near the bank compared to eroded soils sampled well away from the bank. Although soil carbon sequestration would attenuate with distance from the bank, these overall gains of carbon found within water-ponding areas are notable. Organic carbon confers a number of properties to soils that make them more productive, such as improving infiltration rates, cation exchange capacities, and oxygen diffusion rates (Oades 1993).

They also found that with time the amount of available (mineralizable) nitrogen (AN) increased in soils sampled within 10 m upslope of water-ponding

(a)

(b)

FIGURE 5.6. The renewal of productive pastures above water-ponding bank number 6 at a fixed photo point from 1985 (a) to 2001 (b). Photographs courtesy Gary Bastin.

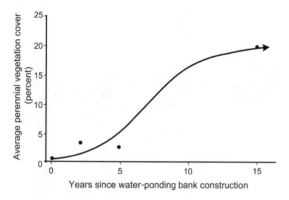

FIGURE 5.7. Vegetation cover increased with time since water-ponding bank construction.

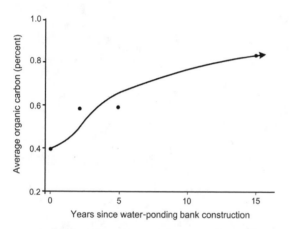

FIGURE 5.8. Soil organic carbon increased with age of water-ponding banks. Organic carbon was measured in the 0 to10 cm soil layer within the 10 m above a bank.

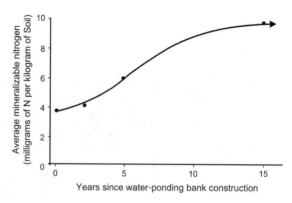

FIGURE 5.9. Available nitrogen increased with water-ponding bank age. Available or minerali-zable nitrogen was measured in the top 10 cm of soil sampled close to banks (within 10 m).

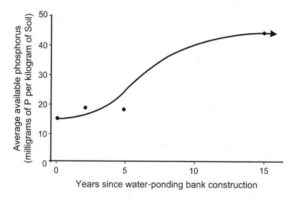

FIGURE 5.10. Available phosphorus increased with water-ponding bank age. Soil phosphorus was measured in samples collected in the top 10 cm within 10 m of banks.

banks (figure 5.9). The concentration of AN had more than doubled near ponding banks compared to never-ponded areas. Average amount of AN in soils collected well above the four ponding banks (i.e., where ponding never occurred) was, on average, only 2.7 milligrams of AN per kilogram of soil. Soil AN in samples collected 10 to 15 m and 25 to 45 m from water-ponding banks were, as expected, intermediate (for brevity, data not shown) between those collected near (less than 10 m) and far (greater than 45 m) from banks. These data indicate a marked improvement in soil fertility.

Another indicator of soil fertility is phosphorus (P). As the age of water-ponding banks increased, they found available soil P increased in soils sampled within 10 m upslope of banks (figure 5.10). The mean concentration of available P in soils collected away from areas that ponded was only 17.8 milligrams of P per kilogram of soil. As expected (data not shown) soil P contents were generally intermediate in samples collected between those near and far from banks.

Mr. Purvis felt confident that his water-ponding bank design was improving soils, because the data demonstrated to him that topsoils (0 to10 cm layer) upslope and near banks had much higher values for three soil fertility indicators compared to topsoils in locations away from banks. These findings suggest that biological processes are most active and persistent close to banks where ponding is more frequent. This is also where grass roots most actively grow. Other biological activities include, for example, litter fall and its processing by invertebrates and decomposition by fungi and microorganisms.

To confirm this greater biological activity upslope of water-ponding banks, one of us (DT) measured soil respiration rates. (See Spain et al. 2009 for methods.) These rates notably increased in the 10 m above banks of increasing age (figure 5.11). Soil respiration rates measured upslope of banks that never ponded were only, on average, 89 mg of carbon dioxide (CO_2) per square meter per hour (averaged across all banks regardless of their age) compared to 226 mg of CO_2 per square meter per hour found just upslope of the oldest bank.

Although these trends in pasture vegetation and soil renewal were clearly successful in an overall sense, as documented by photographs and data (figures 5.6 to 5.11), Mr. Purvis continually evaluated progress in pasture renewal at existing water-ponding banks before building new banks on Woodgreen. From these evaluations he learned the following:

- After water-ponding banks had worked as intended, he no longer needed to sow exotic buffel grass, *Cenchrus ciliaris*, because the abundance of native perennial grass seed had became adequate. This demonstrated a positive biological feedback (from a plant pulse to a seed reserve) within a landscape system. (See figure 2.11 in chapter 2.)
- The soils most responsive to water ponding turned out to have significant swell/shrink properties. With time, the infiltration capacity of these soils improved rapidly because swell/shrink processes improved their capacity to retain water and to sequester organic carbon, thus improving their physical structure. Mr.

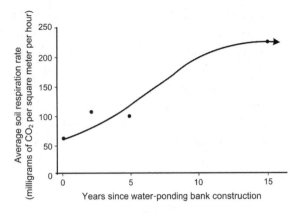

FIGURE 5.11. Soil biological activity (measured as soil respiration rate) increased with time since water-ponding bank construction. Activity was measured within 10 m of banks.

Purvis targeted areas with these soils for new water-ponding banks.

In summary, Mr. Purvis put two key landscape restoration principles into practice. First, he analyzed what had gone wrong (historically) to cause rangeland problems on Woodgreen (principle 1). (See chapter 3.) Second, he identified what critical landscape processes needed to be renewed (principle 2), which in his case were to increase water retention and reduce soil erosion.

Using what is essentially our five-step adaptive landscape restoration procedure, he sequentially applied technologies to retain water and reduce erosion in the landscape by initially using physical technologies (i.e., water-ponding banks) and then applying biological technologies (i.e., establishing perennial pasture plants). He also conservatively stocked rehabilitated paddocks to encourage the recovery of pasture plants so that they more effectively functioned to capture water, protect surfaces from erosion, and cycle nutrients. This also keeps the off-take losses at low levels.

Mr. Purvis is successfully renewing his arid rangelands on Woodgreen. This is verified by a sequence of recent dry years in Central Australia. The drought caused significant cattle deaths on some neighboring stations, but Mr. Purvis continued to produce fat cattle with no mortality. He has had no more rainfall then his neighbors, but his water-ponding banks effectively converted whatever rain fell into productive patches of palatable forage.

This case study confirms that restoration goals are being achieved in an arid rangeland. In our next case study, we document that strategically positioned earthen banks are also effective in restoring semiarid rangelands.

Rangeland, Marra Creek, Western New South Wales

In this case study, our aim is to illustrate how damaged semiarid rangelands can successfully be repaired. As earlier, but more briefly, we use photographs and data to show how water-ponding banks can effectively reclaim areas of bare soil, here referred to as *scalds*, but also described around the world as hardpans, clay pans, or blowouts. The literature on scald reclamation is vast because many different technologies have been tried in an attempt to repair areas with this severe form of soil erosion (e.g., Whisenant 1990). Here we will not try to review this literature but will briefly describe a case of successful scald reclamation on semiarid rangelands located in Eastern Australia where one of us (DT) helped collect data. For details, we refer readers to Thompson (2008).

Study area

Prolonged disturbances in rangelands around the world often cause scalds, a severe form of soil erosion seen as large areas of bare soil (figure 5.12), which remain in this condition for decades, no matter what how good the season or low the grazing pressure. Cunningham (1987) estimated that tens of thousands of square kilometers of rangelands in western New South Wales, Australia, were scalded

by the 1960s. This estimate included about one thousand square kilometers of rangeland in the Marra Creek District located just north of Nyngan, NSW (figure 5.1). The climate in the region is semiarid with summer temperatures often exceeding 40°C. Winter temperatures are often below 0°C. Nonscalded areas grow vegetation that is mostly a shrubland-grassland dominated by saltbush and ephemeral and perennial grasses.

Scalds represent extreme cases of sheet erosion where flows of water and wind have caused the extensive loss of surface soils (figure 5.12). Scalds typically occur on near-level areas, or what are commonly called flats. Prior to being scalded, soils on flats typically have sharp texture boundaries between the A and B horizons. The A horizon erodes because it has a sandy loam texture that is subject to loss by wind and water action when the soils' protective plant cover is damaged. Stripping of the A horizon exposes a hard-setting and dispersive B horizon. Plants have great difficulty establishing and growing on these hard-set soils, hence, scalds tend to stay bare even if conditions improve (e.g., droughts break; grazing pressure is reduced).

Step 1: Setting goals
In 1984, eighteen landholders in the Marra Creek District formed a collaborative team with the Soil Conservation Service of New South Wales to trial water ponding as a technique for repairing their scalded country (Thompson 2008). The team was lead by Mr. Ray Thompson, and they set a goal to return scalds to a productive state where soils have improved soil infiltration, stability, and nutrient-cycling properties, and increased cover of protective and productive vegetation. The landholders in the Marra Creek District desired more productive landscapes to increase the economic viability of their livestock enterprises.

Step 2: Defining the problem
To put landscape restoration principles into practice, the team needed to answer four critical questions

FIGURE 5.12. An extensive scald. Photograph courtesy David Eldridge. (Note in the foreground of this photograph that the scald had been ripped in lines to form shallow troughs and low banks. But, because of the dispersive nature of the exposed soil, these structures rapidly slumped and filled in after rains, so they no longer held water.)

concerning the severity of their scald problem. They conducted analyses to answer these questions. (Each question is followed by its answer.)

1. What caused the scalds to form (e.g., over-grazing by domestic and feral animals) and can these factors be removed or reduced? They found that extensive scalds were caused by long-term excessive total grazing pressure, where livestock and other animals (e.g., feral goats) had exposed soil surfaces to wind and water erosion.

2. Was there any evidence of spontaneous scald recovery, such as expanding "islands" of plants? They found that remnant islands of plants did not significantly expand onto scalded areas even when grazing pressure was controlled or eliminated.

3. What are the rates of infiltration of water into the scalded soils now that they have been eroded down to the B horizon (subsoil)? They found that

infiltration rates of bare subsoils were very low, typically only a few millimeters per hour.

4. What are the dispersivity, sodicity, and shrink/swell properties of these exposed subsoils? (These three soil properties are defined in the glossary.) Scalded soils were found to be sodic and highly dispersive, but fortunately were found to shrink (crack) on drying after being wet.

Steps 3 and 4: Designing solutions and
      applying technologies

To achieve their goals to increase the cover of vegetation, and improve soil properties on scalds, the team designed and sequentially applied two basic restoration treatments:

1. First, physical technologies were applied by building water-ponding banks that would function to retain water within the scalded areas after signif-

icant rains (figure 5.13). Machinery was used to construct ponds with the following design attributes (Thompson 2008):

- Earth banks were precisely constructed (by laser-contouring) to be 0.5 meters high and 2 meters wide at the base.
- Maximum size of a ponded area was about 4,000 square meters (0.4 hectares), because early trials demonstrated that ponds with larger surface areas tended to have wind-induced wave action that caused bank failures (breaks).
- On scalds with a slight slope (up to 0.4 percent) ponds were constructed to be U-shaped, but on flats they were elliptical (figure 5.14).
- Soil surfaces within elliptical or U-shaped areas were not treated (ripped, tilled) so that any residual soil biological properties and any remnant vegetation would not be disturbed.

- Overall, the aim was to pond water to a maximum depth of 0.1 meters, which would establish native upland plant species rather than wetland species.

2. Second, the team initiated biological processes by

- Encouraging the establishment of native rangeland plant species within ponds by controlling the maximum depth of water, as noted earlier.
- Sowing seeds of species known to be adapted to growing on hard-setting soils. In some ponds, sowing was not necessary because seeds from natural vegetation were produced by plants on remnant islands of vegetation within or near the pond.
- Controlling grazing disturbances to as little as possible in the early stages of rehabilitation and then carefully regulating grazing in later

FIGURE 5.13. An example of a water-ponding bank constructed on a scald. The bank is successfully retaining water after a rain event. Photograph courtesy David Eldridge.

FIGURE 5.14. A set or network of elliptical and U-shaped water-ponding banks constructed to capture and retain rain water over a large scald. Photograph courtesy Ray Thompson.

years to encourage vegetation establishment on the newly treated areas; this is an important restoration principle.

Over time these biological processes would be expected to largely replace the function of physical processes improved by water-ponding banks. (See figure 3.1 in chapter 3.)

Step 5: Monitoring and assessing trends
To evaluate the effectiveness of water ponding as a technique to reclaim scalds, the team observed and measured the following on water-ponding areas (Thompson 2008):

1. Were ponds holding water after major rain events (water-ponding banks were checked for breaks or breaches)?
2. Was vegetation establishing and forming a protective surface cover on ponded areas (photographs at fixed points were taken and vege-

tation composition and cover were estimated)?
3. Were soils improving (surfaces were examined to see if they were becoming soft and crumbly and if they were cracking due to swell/shrink properties; if so, cracks wider than 20 mm indicated responsive soils)?
4. Were soil-surface conditions improving as measured by surface stability, infiltration, and nutrient-cycling indices? (Monitored by one of us [DT] using methods described in chapter 14.)

After monitoring over a number of years, which included seasons with major rains, the team evaluated indicators for meaningful signs and trends, and found the following:

- No breaches along the banks, so that water-ponding banks were effectively holding water over extensive areas (plate 5.3).
- A mix of vegetation had quickly established within water-ponding areas (figure 5.15).

(a)

(b)

FIGURE 5.15. A severely scalded area just prior to water-ponding bank construction (a), and eleven months later (b). Photographs courtesy Ray Thompson.

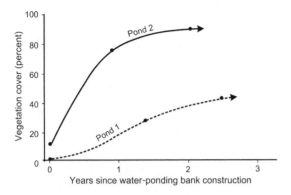

FIGURE 5.16. Vegetation recovery after two years on two ponds located in the same scalded area. Recovery was faster on pond 2, which initially had a 12 percent cover of remnant vegetation compared to a totally bare pond 1. Data courtesy of Ray Thompson.

- Cover of vegetation differed on ponds because of their initial conditions; for example, one pond initially had no cover and the other 12 percent cover (figure 5.16).
- Soils were developing favorable properties because of water ponding. This was evident from soil surfaces becoming friable and cracked instead of hard setting. (See figure 5.5.) There were also improvements in soil-surface stability, infiltration, and nutrient-cycling indices but, for brevity, these data are not presented here.

Based on their observations and assessment of trends in these indicators, the team concluded that the application of water-ponding technologies was successfully repairing scalds within their semiarid rangelands (Thompson 2008).

Further Thoughts

We want to emphasize that by putting water-ponding technologies into practice, damaged rangelands are effectively restored. This is because water ponding is a technology that applies our principle 2: it sequentially and effectively repairs physical and then biological landscape processes (i.e., banks are built to retain water, and then seeds germinate in this moist soil to eventually form a protective and productive cover of vegetation). To support this point, we note that water-ponding technologies have successfully retained water and grown useful vegetation in other arid rangelands. For example, in arid Central Australia on Bond Springs Station, Richards and Walsh (2008) found that, after nearly forty years, banks constructed to pond water were the only persistent and effective treatment compared to other treatments such as contour plowing, pitting, disking, and ripping.

We finally note that Mr. Thompson and his team, working in collaboration with a land reclamation group in the Marra Creek District near Nyngan in New South Wales, Australia, have surveyed and constructed over 50,000 water ponds on scalded areas. Over the years they have fine-tuned their water-ponding technology (Thompson 2008). Water-ponding technologies are now being applied to scalded rangelands in the United States, China, France, Israel, and a number of African countries.

PART III

Scenarios for Restoring Landscapes: Mine Sites, Rangelands, Farmlands, and Roadsides

In part 3, we describe how our five-step adaptive restoration procedure success-fully repairs damage to a variety of landscapes. We illustrate how this procedure puts principles into practice to (1) renew cleared farmlands (rural and suburban); (2) repair disturbed roadsides; (3) reclaim other mine sites (waste-rock dumps, tail-ings storage facilities, open-cut spoil heaps); and (4) restore other damaged rangelands (shrubby and eroded). Our goal is to illustrate how our procedure generic-ically applies over a very wide range of climates and types of vegetation and soils.

In the following chapters we use what we call a scenario approach because we want to show how to apply our five-step adaptive procedure to a range of cir-cumstances. Such a generalized approach is useful because options can be explored, and there are so few completely documented examples, or case studies, available as public documents.

Scenarios are narratives about likely future developments in different situations. They are not predictions about what will actually happen. We urge readers to keep in mind that the scenarios we present in the chapters of part 3 are not actual land-scape restoration case studies, but they are plausible restoration problems. Although our scenarios are constructs, the outcomes we describe are highly likely because they are based on our many years of experience in studying how to successfully restore different types of landscapes (e.g., Tongway 1995; Ludwig and Tongway 1996; Tongway and Ludwig 2007). The data trends we present for dif-ferent restoration indicators have been generalized for the purpose of our scenarios, but trends are based on data we have collected, or that have been collected in stud-ies by restoration practitioners and students we have worked with over the years.

From many possible scenarios, we have chosen seven types of landscape damage faced by restoration practitioners. Each type of damage is presented in a separate chapter. We describe how adaptive landscape restoration applies to three types of damage on mined lands (chapters 6 to 8), on two types of grazed lands (chapters 9 and 10), on former pasturelands (chapter 11), and on roadsides (chapter 12). In presenting scenarios in these chapters, we extensively use a bul-let (dot-point) format. This is a concise way to emphasize the steps and principles involved in solving different landscape restoration problems.

In most of the seven scenarios, we describe situations where restoration trends are "not OK." (See figure 1.1 in chapter 1.) In this situation we activate the adaptive learning loop shown in figure 1.1 to illustrate how restoration trends are improved. We also include scenarios that are in the early stages of landscape rehabilitation. In these examples, early trends for indicators are positive, and restoration is following our five-step adaptive procedure and its underlying principles, so that we are confident that positive trends will continue toward a declaration of successful landscape restoration.

Chapter 6

Restoration of Mine-Site Waste-Rock Dumps

In this first scenario chapter, we describe how our five-step adaptive procedure applies to the restoration of waste-rock landscapes created by *hard-rock* mining. Our aim is to illustrate that by putting our procedure and its principles into practice, these highly disturbed landscapes can be successfully restored. We use the term hard-rock mining to cover all those operations around the globe that extract valuable minerals such as gold, silver, lead, zinc, copper, nickel, tin, and uranium from ore bodies composed of rock having low weathering rates. Within these ore bodies, waste rock is the material that has no commercial mineral content.

Depending on the depth at which these mineral-bearing rocks are located, hard-rock mining operations typically disturb landscapes by creating large dumps of waste rock using two types of excavation methods:

1. If close to the surface, open-pit excavations are used to mine minerals (figure 6.1). Waste rock and regolith (the layer of heterogeneous materials covering an ore body) are removed from the pit, dumped in piles, and pushed into plateau-like waste dumps (plate 6.1), which can cover hundreds of hectares. These waste-rock dumps are usually located near the pit to minimize haul distances. Mineral extraction processes also produce finely ground rock materials or tailings, which are also stored in facilities located near the pit. We will describe how restoration practitioners (RPs) can put principles into practice to restore land covered by tailings in the next chapter.

2. If the mineral-bearing rocks are located deep underground, mine shaft or tunnel excavations are used to extract them. Deep underground mining produces a little waste-rock material and larger amounts of tailings. Backfilling of underground tunnels, or exhausted pits, with waste materials can be done in some cases, but this option is infrequently used because it closes off future mining options. Above-ground storage of waste rock in dumps is usually the preferred option.

To restore landscapes from waste-rock dumps, RPs face many challenges. They can successfully meet these challenges by applying our five-step adaptive procedure and by putting landscape restoration principles into practice. We state this with confidence because of our experiences, and those of colleagues, on restoring waste-rock dumps in mine sites located around the globe (e.g., Loch 1997; Ludwig et al. 2003; Tongway and Ludwig 2006; Vasey et al. 2000). We have used these experiences to generalize the findings we present in this scenario chapter.

Setting the Scene

This scenario deals specifically with restoring waste-rock dumps (plate 6.1) created by operations at a typical open-pit, hard-rock mine site (figure 6.1). For this scenario, we assume the following:

- The climate at the mine site is, overall, moderate and favorable for plant growth. Annual rainfall is assumed to be about 700 mm, and temperatures in summer rarely exceed 40°C. In winter, temperatures occasionally drop below 0°C. The scenario mine site is in a temperate, but mild, climatic setting.
- Stockpiles of topsoil, created when stripping

the vegetation and soils away from the area being mined, are relatively small compared with the amount of topsoil needed to cover restored areas; this shortage of topsoil is typical for most hard-rock mines around the world.
- The waste-rock dump being restored is a heterogeneous mixture of mined materials (plate 6.2) composed of hard rock, which is resistant to weathering, and regolith, which readily weathers; globally, this type of waste-rock dump is common on hard-rock mines.

Step 1: Setting goals
In this scenario, the stakeholders in the land being mined (e.g., site lease holders, government regulators,

FIGURE 6.1. A typical deep open-pit created by a hard-rock mining operation. In this case, about 50 m of regolith (mostly oxidized broken rock or oxide) is located above the main ore body.

local community groups) required that the RP achieve four goals: (1) Waste-rock/regolith piles must be reshaped into stable landforms that blend with surrounding natural landscape features, such as plateaus, rolling hills, and valleys. (2) Reshaped landforms must discharge less than a specified amount of runoff and sediment from the mine site, which is set by the regulators. (3) Any acidic drainage must be contained on-site, along with any other contaminants or pollutants formed and mobilized by water leaching through waste rock. And, (4) vegetation established on these landforms must protect surfaces from dust generation and must also be adapted to surviving in the long term without management inputs (i.e., be self-sustaining).

Step 2: Defining the problem

To achieve the four goals, the RP worked with experts to critically analyze various strengths and limitations of the waste-rock/regolith materials on the dump, along with other factors likely to affect land forming and revegetation:

- Because of the large piles of waste rock/regolith (plate 6.1), mine-site engineers informed the RP that reshaping dumps into stable landforms would be costly.
- To minimize mining costs, site engineers simply dumped waste-rock/regolith materials with differing properties into mixed piles (plate 6.2) rather than sorting them into different stockpiles, which could then be used later, for example, to cap spots of acid-forming waste rock.
- Geochemists found that rates of weathering and erodibility of different waste-rock/regolith materials (plate 6.2) varied from very low to very high.
- Geochemists also found that some waste-rock/regolith materials contained significant concentrations of acid-forming minerals, which could dissolve heavy metals and release them off-site into aquifers.
- To estimate the potential for storms to trigger large runoff and erosion events, the RP exam-

ined rainfall records. Although annual rainfall was only 700 mm, the record showed that high-intensity rainstorms occurred in some years.

- The RP knew that the amount of available topsoil was limited, and unfortunately also found that this topsoil was prone to dispersion and slaking. See *slake test* and *slumping* in the glossary. (Topsoil in this context means soil derived from stripping regolith surfaces down to about 0.3 m before open-cut mining; it does not imply garden soils.)
- Because most topsoil was stockpiled for decades, the RP found that opportunities to use biologically "fresh" topsoil (i.e., with viable seeds and soil organisms such as termites, ants, earthworms, fungi, and bacteria) was limited.
- Soil biologists found that soluble salt concentrations in some of the waste-rock/regolith materials were high enough to threaten the success of establishing vegetation, and leaching these salts could create off-site contamination problems.
- To explore constraints to growing vegetation on the waste-rock/regolith materials, the RP and researchers conducted field experiments and examined other rehabilitated sites. They found that plant species composition and life-forms on these sites were not typical of species occurring in surrounding natural landscapes, but the plants present were functioning to retain flows of water and prevent erosion and were usefully contributing organic materials (e.g., litter) to nutrient cycles and to soil formation.

Steps 3 and 4: Designing solutions and applying technologies

Having analyzed the various strengths, problems, and challenges with restoring waste-rock dumps, the RP worked with experts to design and apply feasible solutions and technologies. They first worked on designing and building stable landforms (i.e., applying physical restoration technologies). (See figure 3.1 and principle 2 in chapter 3.) To account for the potential erodibility of waste-rock/regolith and topsoil materials, the RP worked with landscape evolution modelers to optimize landform designs:

- Where slopes were required, they designed and constructed landforms to have concave structures such as bank-and-troughs running along contours. Previous models and applications had demonstrated the long-term effectiveness of such concave structures to retain water and sediments on-site. See Loch (1997), Willgoose and Riley (1998), and Hollingsworth (2010) for landform designs on mine sites.
- Where plateau-like areas could be formed, they constructed surfaces into small, internally draining basins (a form of a *store and slow-release water* structure). These structures (1) greatly enhanced vegetation establish-

ment and growth from small rainfall events, (2) minimized the potential for on-site wind and water erosion, and (3) reduced runoff rate and volume.

After these landforms were constructed, the RP worked with site engineers and machinery operators to physically prepare sites for subsequent revegetation:

- To create surfaces favorable for plants, they first covered reshaped surfaces with regolith, which was mostly coarse broken rock (figure 6.2). Then they applied available topsoil, and mulch and woody debris (figure 6.3), because these surface coverings function to (1) prevent direct raindrop impact onto the soil surface, (2) obstruct and deenergize overland flows, (3) enhance infiltration, and (4) reduce the potential for wind to blow dust off-site.

FIGURE 6.2. An example of a waste-rock dump that has been shaped into a plateau-like landscape and covered with coarse regolith materials. Note the approaching rainstorm.

Figure 6.3. A rehabilitated mine site where woody debris has been applied to the surface of a reshaped waste-rock dump.

- To reduce the potential for valuable topsoil to be lost during storms, they roughened reshaped surfaces by tilling (shallow ripping) on the contour. As a physical process, tilling also "keys-in" the topsoil with deeper substrates; this avoids the layering of materials into discrete bands, which may form barriers to infiltration and may promote landslip erosion.
- To prevent damage to plants by any spots of acid-forming materials, they promptly and permanently applied a capping of materials (e.g., non-shrink/swell clays) to these spots. The aim of this capping was to form a barrier to water and oxygen, which slows acid formation to the lowest possible rate and prevents

any above- or belowground flows of this acidic water off-site.

After physically preparing rehabilitation site surfaces, the RP worked with botanists and ecologists to plant vegetation to form self-sustaining ecosystems:

- To increase the likelihood that vegetation would be suited to the local climate, they collected seeds from local natural vegetation and sowed these seeds to augment any viable seeds in applied topsoil.
- To enhance germination and establishment, they timed sowing to take advantage of expected rains.
- To ensure that vegetation included *framework* plants, they specifically collected seeds from these key species and grew them as tube-stock in nurseries. Framework plants are those species known to provide important ecosystem benefits, such as goods and services for local people and special habitats required by animals.
- To take advantage of run-on areas created in the reformed landscape, such as internally draining basins, they strategically planted tree seedlings (as tube-stock) within these areas, rather than in uniform rows as in plantations.
- To promote the growth of seedlings, they added only a small amount of nitrogen and phosphorus fertilizer. They used fertilizer applications similar to nutrient concentrations found on natural reference sites. They also knew from previous experience that heavy applications of fertilizer encouraged weeds on rehabilitated mine sites.
- To control colonizing and aggressive weeds, they applied herbicides.

Step 5: Monitoring and assessing trends
To evaluate restoration trends as soon as possible after sequentially applying physical and biological restoration technologies, the RP immediately initiated a protocol to monitor indicators reflecting landscape

processes so that, if necessary, adjustments to technologies could be made to improve restoration trends. The RP selected a number of key indicators to monitor:

1. Signs of erosion, indicating unstable surfaces on recently constructed landscapes. The RP observed and measured the following:

- The extent of any lateral flows of runoff along troughs created along slope contours by deep-ripping, which forms a system of banks with concave troughs designed to retain water.
- Any water pooling in troughs along contours; such pools are where water is likely to overflow banks and initiate rills and gullies.
- Any new rills on reformed and revegetated slopes; if found, the RP monitored their width and depth.
- Any areas of fresh alluvium found at the foot of slopes; if found, the RP identified their source and monitored their size.
- Any gullies; if found, the RP measured their size (depth, width) and located their source.

By monitoring these signs of erosion, the RP could initiate repairs (e.g., fix overflow points along banks) before they became serious (e.g., deep gullies) and costly.

2. Water quality, indicating potential problems with off-site pollution. The RP regularly monitored the following:

- Surface waters and water extracted from aquifers, both on- and off-site, were monitored for contaminants, which if found signaled the immediate need for remedial actions.
- Hot spots of acid-forming waste rock, which if found immediately required the application of a thicker capping to better isolate these spots from water and oxygen.

3. Vegetation establishment and long-term survival, indicating if a self-sustaining ecosystem is developing. The RP measured indicators along permanently positioned transects oriented downslope, which are called *gradsects* (Gillison and Brewer 1985), using standard procedures. (See chapter 13.) These procedures include observations and measurements on these features:

- The position of vegetation and bare soil patches; the RP repaired any large gaps in vegetation as soon as possible.
- Sizes of vegetation patches (horizontal and vertical dimensions); the RP assessed rates of patch growth and their role in retaining water and soil, and providing habitats for animals.
- The species composition of the establishing vegetation; the RP evaluated whether vegetation trends were toward those expected from natural reference sites.
- The presence and abundance of exotic weed species; the RP determined if any control treatments were needed. If aggressive weeds were found, they were quickly controlled, whereas weeds of low persistence were given a lower priority.
- Amounts of remaining mulch and woody debris, and the accumulation of litter from the developing vegetation; the RP assessed whether these organic materials were being decomposed and incorporated into the soil by fungi or soil invertebrates. If they were being washed away, the RP took corrective measures to stop excessive runoff.

For this scenario on restoring waste-rock dumps, the RP generally monitored progress at annual intervals. However, some indicators, such as water quality, were monitored more frequently, especially if a major disturbance (severe storm) hit the site and repairs were needed.

The RP continually analyzed trends in indicators and, after more than ten years, found a number of positive developments. For example, the growth of trees was toward that expected from unmined reference sites (figure 6.4a), although many more years will be needed for trees to reach full size. (Note, although these trends are generalized for this scenario, they are based on data trajectories reported in Ludwig et al. 2003.) The deep rip-lines in the reshaped landforms were very resistant to in-filling over time (figure 6.4b) and remained effective in retaining water, topsoil, and litter produced by vegetation leading to formation of new soil. This means that on rehabilitated waste rock the need for vegetation to take over the function of resource retention from collapsing rip-lines is less critical than for other mines (e.g., bauxite) where rip-lines are created in less robust material.

The RP found that trends in the percentage composition of the trees were generally toward an increase in long-lived *Eucalyptus* species and a decrease in short-lived *Acacia* species (figure 6.5) as would be expected for the development of self-sus-

taining ecosystems similar to those found on nearby reference sites. However, the trend in establishment of other woodland species remained below that expected, and the RP took remedial actions by planting missing tree species (grown in containers from locally sourced seeds) into the rehabilitating site. The RP also found from tests of soil properties (e.g., salinity) that some areas of soil were not suitable for some plant species. The RP consulted with stakeholders to approve the cost of treating the soil to render it more favorable for local plants.

To assess the development of habitats for fauna, the RP calculated an index of habitat complexity from vegetation structure measurements (e.g., horizontal and vertical tree canopy dimensions) and other site attributes. (See chapter 16.) The RP found that habitat complexity rapidly increased toward that expected from natural unmined sites (figure 6.6). However, special habitats such as tree hollows take much longer than ten years to develop, and absence of hollows at this relatively short time scale was not interpreted as restoration failure. To fully assess faunal habitat restoration, the RP continued to monitor vegetation structure and animal diversity.

To evaluate soil development on rehabilitated sites, the RP examined trends in soil-surface condition indicators relative to those expected from natural

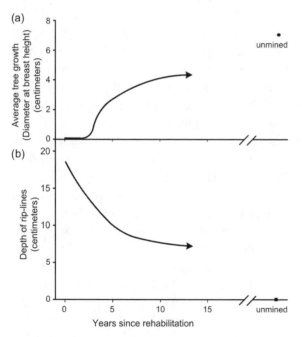

FIGURE 6.4. (a) Trend in tree growth (mean DBH) on rehabilitated mine sites was toward that expected for unmined sites. (b) Rip-lines created in surface materials remained 7 cm deep after 10 years (no rip-lines on unmined sites).

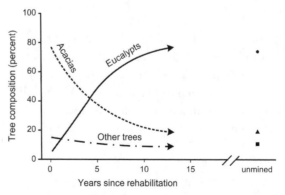

FIGURE 6.5. Trends in tree composition on rehabilitation sites toward those expected from nearby unmined reference sites for acacias (square), eucalypts (circle), and "other" tree species (triangle).

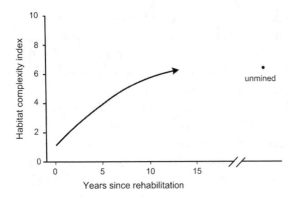

Figure 6.6. The trend in the habitat complexity index is toward the level expected for a natural unmined landscape.

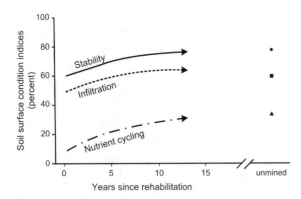

Figure 6.7. Trends in soil-surface condition indicators on rehabilitation sites: stability (circle), infiltration (square), and nutrient cycling (triangle) are toward values on unmined sites.

unmined sites. The RP found that surface stability and infiltration indices were initially high (figure 6.7), and remained so because the surfaces formed by the physical ripping of coarse waste-rock/regolith materials were inherently stable and readily took in water. The nutrient-cycling index steadily increased

as vegetation developed, because the biological processes of litter fall and its decomposition became prominent.

All of the above monitoring data were evaluated by the RP for progress toward the stated goals. After twenty years, the RP found that all trends were OK, and all goals were being achieved. The RP presented these findings to stakeholders and regulators who concluded that the restoration of waste-rock dumps on their mine site was succeeding.

Further Thoughts

We end chapter 6 by emphasizing the importance of not only applying the five-step restoration procedure and putting principles into practice, but also how essential it is to continue monitoring, because if negative trends in indicators are detected at early stages of rehabilitation, potential problems can be fixed at low cost. Prompt provision of monitoring data to the manager enables them to apply the *trends OK/trends not OK* test. (See figure 1.1 in chapter 1.) If problems are found with some indicator trends, then the answer to the question, is the trend OK? is no. In this case, the RP can analyze underlying causes of the problem and design fixes to implement (i.e., apply the adaptive learning loop; figure 1.1). For example, if the RP detects signs of erosion, such as rills on a reformed slope (figure 6.8), the cause can be isolated (e.g., a broken bank) and correcting actions taken. In this case, the bank can be repaired and the movement of sediments down the rill can be arrested inexpensively by applying, for example, locally available hay mulch (figure 6.9). Following this treatment, additional monitoring allows the RP to check on its effectiveness (plate 6.3).

FIGURE 6.8. An example of a slope that has eroded into rills, with water ponding at the base of the slope rather than infiltrating into the slope to benefit plants.

FIGURE 6.9. An example of the application of hay mulch to trap sediments on an eroding slope (see figure 6.8). All sediments were being trapped in the upper 40 cm of the applied mulch.

Chapter 7

Restoration of Mine-Site Tailings Storage Facilities

In chapter 6 we described the challenges restoration practitioners (RPs) face when restoring landscapes from waste-rock dumps. Our aim in this chapter is to present what is perhaps a greater challenge: restoring landscapes on *tailings storage facilities* (TSFs), often called tailings ponds, at mine sites; these TSFs occur around the world. Tailings are the finely ground materials remaining after mining and processing rocks to extract precious metals such as gold, silver, lead, zinc, and copper. After processing, tailings still contain varying amounts of heavy metals and chemical contaminants, which have adverse reactive chemical, physical, and biological properties that present major restoration problems and can threaten human health.

Globally, tailings are stored within constructed aboveground embankments (figure 7.1). Although

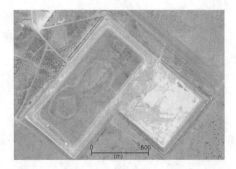

FIGURE 7.1. An aerial view of tailings storage facilities constructed as rectangular ponds with steep sides to reduce their landscape footprint. Photograph courtesy of Harley Lacy.

wall construction varies, TSFs are typically initiated by building a low, encompassing starter wall of earth and rock. Tailings placement usually begins just inside the outer walls and proceeds toward the center. When this initial pond is nearly full, the dry tailings from around the edges of the pond are reshaped to form a new higher wall. This construction process is repeated over time to create a series of lifts where higher sections of the wall are stepped in to form a truncated pyramid shape (figure 7.2), which also provides road access to the top of the tailings pond.

TSFs are typically filled by pumping finely ground tailings in pipes as slurry from processing plants to their containment within embankments. Some newer processes pump tailings as a paste after partial dewatering. Because of the logistics and economics of pumping slurry or paste, TSFs are usually located near mineral extraction plants. TSFs may cover very extensive areas; for example, they cover about 500 km^2 in the Witwatersrand Region of South Africa.

Restoration of TSFs usually does not begin until after tailings deposition operations have ceased, and after tailings have dried sufficiently to bear the load of vehicle traffic carrying cover materials. Because of the way they are constructed, restoring TSFs involves rehabilitating (1) the relatively flat, plateau-like tops of the ponds, and (2) the sloping embankments or walls forming the pond flanks.

There are two different ways of restoring TSFs, depending on whether ore bodies are mined by sur-

FIGURE 7.2. Stepped walls of a tailings storage facility built from dry tailings and covered with nonreactive waste rock.

face open-pit operations or by tunneling underground. Around the world, open-pit mining operations typically produce sufficient waste-rock and regolith materials to completely cover tailings as part of the landscape restoration procedure (e.g., figure 7.3; plate 7.1). In this case we recommend RPs put into practice our five-step adaptive procedure and its principles. We described these practices in our chapter 6 scenario where, to retain water and promote vegetation on waste-rock dump slopes (figure 7.4), for example, RPs constructed bank-and-trough systems along contours (figure 7.5). Because these same landscape restoration procedures and technologies also apply to reclaiming waste-rock/regolith-covered TSFs, we will not repeat these procedures here.

Here we describe how RPs can restore landscapes on TSFs located on mines using underground tun-

FIGURE 7.3. The entire top of a TSF is being covered with earth and waste rock. Photograph courtesy of Harley Lacy.

Figure 7.4. An example of vegetation established on a gently sloping TSF wall covered with earth and waste rock in a semiarid environment.

Figure 7.5. An example of a bank-and-trough system constructed on a low-angled TSF wall. Natural landforms can be seen in the background. Photograph courtesy of Roger Potts.

neling operations, which produce insufficient topsoil and waste-rock/regolith materials to completely cover tailings. In addition, most tailings produced by underground mining operations remain uncovered, because if more efficient mineral extraction processes are developed, these tailings have the potential to be reprocessed at low expense. For these reasons, TSFs on underground mines are very rarely restored by completely covering tailings with waste-rock/regolith materials. For example, the extensive TSFs in the Witwatersrand area of South Africa, noted earlier, are uncovered. The problem is that when raw tailings are not covered to protect them from exposure to wind and rain, RPs must deal with a number of difficult technical problems such as TSF wall erosion (figure 7.6). Along with mine stakeholders, RPs must also deal with social issues, such as dust pollution and off-site heavy metal contamination.

Problems with restoring sustainable landscapes on uncovered TSFs are faced by RPs on mine sites around the world such as those in Australia, Europe, North and South America, Africa, and Asia. Because restoring uncovered TSFs is so common, yet so difficult, we have generalized our scenario for this chapter. We describe how, by applying landscape principles and an adaptive restoration procedure, RPs can overcome a number of difficulties to successfully restore uncovered TSFs. We do this with confidence because of our experiences in working with colleagues on restoring TSFs, especially in Australia and South Africa (e.g., Lacy and Barnes 2006; Williams and Kline 2006; Weiersbye and Witkowski 2007). These experiences build on those of others on rehabilitating TSFs in North America. (See, for example, Peters 1984.) For emphasis and brevity, we use bullet points in this scenario.

FIGURE 7.6. Overflow from the top of a TSF (at top right) has eroded the wall, carrying tailings off the site.

Setting the Scene

As noted, this scenario specifically deals with restoring landscapes on TSFs on sites using deep shaft mining operations where only very small quantities of waste-rock, regolith, and topsoil materials are produced. For this scenario we assume the following setting:

- Tailings on the plateau-like top of the TSF are uncovered, and those on steeper slopes of stepped sides may be partially covered with any available waste-rock/regolith/topsoil materials. (See figure 7.2.)
- The climate is moderate and favorable for the growth of plants adapted to warm, rainy summers and mild, dry winters.
- The RP understands how the TSF is positioned in relation to previous land surfaces and subsurface drainage systems.

Step 1: Setting goals

Mine-site stakeholders and the RP agreed that their general objective was to convert the TSF into a landscape that meets safety, environmental stability, and aesthetic acceptability criteria. Because tailings are potentially toxic and have a very fine, noncohesive texture, this means they needed to minimize the risk of off-site contamination. The stakeholders set a number of specific goals for the RP:

- To reduce the potential for tailings dust to blow off-site, the RP needed to establish a self-sustaining cover of, ideally, native vegetation.
- To limit the flow of any contaminated water off-site to very low (agreed to) levels, the RP needed to establish plants that would function to minimize deep percolation through tailings; this could be achieved by having plants transpire water that might percolate through tailings and emerge as contaminated base flow.
- To reduce the risk of contaminating water by its percolation through acidic hot spots, the RP needed to find any areas of highly reactive

pyritic materials where acidic percolation could dissolve out contaminates such as heavy metals, which could then potentially flow off-site. If found, these hot spots could be treated with limestone or dolomite to neutralize any acid formed by pyrite oxidation. However, the large bulk of the tailings would retain their original properties.

Step 2: Defining the problem

The RP analyzed the range of issues in restoring uncovered TSFs within the context of the broad landscape setting. A number of challenges became evident.

Potential toxicity

Because of differences in the rocks being mined, tailings vary greatly in composition over the life of the mine. This is evident in the color bands seen in TSF walls (plate 7.2). Because tailings contain varying concentrations of toxic heavy metals and pyritic materials, they oxidize to form sulphuric acid when water and oxygen are available; this acid dissolves metals held in tailings and potentially releases them off-site. The RP had samples of tailings collected from the site analyzed for the following:

- General physical and chemical properties (e.g., particle size; and pyritic, toxic metal, and residual chemical content)
- Specific potential for pyrite oxidation, including their oxidation rate, so that technologies could be designed to minimize the potential risk of any plastic flows and drainage (both surface and subterranean) from the TSF

Wall erosion and stability

Until the RP establishes a protective cover of vegetation, uncovered walls of the TSF are exposed to the erosive forces of wind and water (figure 7.6). Furthermore, the core of a tailings pond remains moist for many years after the addition of slurry has ceased, but the external walls of the TSF can dry rapidly. As external walls dry, their brittleness, and the mass of still-moist core of tailings behind it, can cause faults, which may release plastic flows of tailings through the wall that spread slowly over the external landscape and plumes of contaminated water that percolated out from the base of the wall and flow off-site.

The RP knew from observations that TSF wall failures are very difficult to predict and control and that off-site leakages can pose significant short- and long-term environmental and health threats to human populations in the region. The RP recognized an additional problem: many TSFs are located on natural drainage lines because these areas are not traditionally used for agriculture or horticulture. This meant that any leakage from the base of the TSF directly enters surface or subsurface aquifers and, if toxic, causes off-site contamination.

Dust pollution

Because tailings are composed of finely ground rock material about 30 microns in size, which are fine enough to be transported long distances as dust in wind, the RP had a major problem with the uncovered TSF: every storm has the potential to carry hazardous materials off-site.

Soil formation

Because of their finely ground nature, tailings superficially resemble fine sandy or silty soils, but the RP knew tailings do not have the properties of real soils:

- Tailings do not initially contain actual clays, although there may be clay-sized rock particles present. Unlike true clays, dry tailings have moderate infiltration rates, but low water-holding capacities.
- Tailings have poor *cation exchange capacity* (CEC), which is an important property of real soil that is associated with clays. As the name implies, CEC is the capacity of the soil to act as a nutrient reserve that reversibly sorb cations such as ammonium, potassium, and calcium; these cations are essential for plant growth.

- Initially on TSFs, there are few habitats suitable for soil biota ranging from invertebrates to fungi; the lack of these organisms delays soil-forming processes.

Establishing vegetation

The RP analyzed raw tailings and found them deficient in nearly all critical plant nutrients, which makes raw tailings a poor medium for establishing plants. Although clays and other soil components eventually form from tailings by natural soil-formation processes, the RP knew that the timescale is far too slow; the goal is to quickly establish vegetation to control water and wind erosion.

Steps 3 and 4: Designing solutions and applying technologies

The RP faced major challenges in dealing with all of the above problems and decided to systematically explore different designs and technologies to rehabilitate the relatively flat tops of the tailings ponds and the steeper sides (walls) of the TSF.

After considering different options, the RP designed a procedure to establish a cover of grasses on raw tailings by applying water (irrigate), nutrients (fertilize), and ameliorants (lime or dolomite). The RP then sowed a mix of grass seeds for species known from field trials to be adapted to the climatic setting and to growing in raw tailings being limed, fertilized, and irrigated. Although irrigating, liming, and fertilizing tailings is costly and requires constant maintenance (Haagner 2009), the RP's goal to quickly cover the TSF top with grasses to prevent off-site dust contamination was deemed paramount. A thick grass cover also transpires incident rainfall to reduce deep drainage through tailings and potential seepage from the base of the TSF wall.

Step 5: Monitoring and assessing trends

After applying the above technologies to establish a TSF landscape covered with irrigated grasses, the RP immediately began monitoring a number of restoration indicators. For example, the RP annually measured the grass cover and also the composition of the species that grew on rehabilitated sites, on both the tops and sides (walls) of the TSF. The RP also checked for signs of soil erosion on these sites after storm events, and for any signs of contaminated water leaking from the base of TSF walls. See chapter 6 for other examples of landscape restoration indicators.

After eight years, the RP found that restoration trends on the tops and walls of the rehabilitated TSF were positive. However, because of the constant maintenance demands and high costs of supplying irrigation, lime, and fertilizers to extensive areas of rehabilitation, the RP (in consultation with stakeholders) decided to "turn off the tap."

Turning the tap off

When irrigation and fertilization had ceased for only two years, the RP observed a number of severe problems:

- Because the grass root systems had developed to expect about twice the natural rainfall, grass cover on the top and walls of the rehabilitated TSF rapidly declined from high eight-year values of 85 percent obtained under irrigation and fertilization down to about half these maximums in year ten (figure 7.7). The natu-

ral grass cover on reference sites was considerably lower than the artificially maintained maximums.

- Because of lower grass cover, the RP found that soil surfaces were forming physical crusts that readily shed rainwater (figure 7.8) and signs of surface erosion were becoming evident (pedestals, rills).
- With less grass cover to transpire water after major storm events, water was leaching through the tailings and leaking from the base of the TSF wall and entering local drainage networks. This seepage was found to be contaminated with toxic heavy metals.

The RP learned that these problems have also been found on other TSFs where grasses were grown under irrigation and fertilization on uncovered tailings.

Given the seriousness of the above problems, the RP informed the mine-site stakeholders that rehabilitation trends were clearly not OK. (See figure 1.1.)

Trends Not OK: Follow the Adaptive Learning Loop

In a workshop with the stakeholders, the RP worked through the *adaptive learning loop* (figure 1.1).

Step 1: Resetting goals

First, the RP and stakeholders reexamined their goals and agreed that, while it remained very important to quickly grow vegetation on TSF tops and walls to prevent wind and water erosion, their new aims were as follows:

- To establish long-lived savanna and woodland vegetation surface covers that depend only on natural rainfall, rather than the fast-growing grass vegetation initially trialed, which was dependent on irrigation and fertilization.

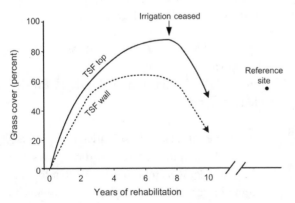

FIGURE 7.7. The mean trends in grass cover on rehabilitating TSF tops and walls. As a reference point, grass cover was also measured on nearby natural grassland sites.

FIGURE 7.8. Signs of surface erosion and compact physical crusts are evident between grass clumps along a transect being monitored on the walls of a rehabilitated TSF.

- To devise a strategy to contain on-site any leakage from the base of the rehabilitated TSF walls.

Although establishing savanna and woodland vegetation on TSF tops and walls meant incurring the costs of reworking raw tailings and adjusting outer wall slopes, these initial costs were expected to be offset in a few years by removal of the ongoing costs of the irrigation and fertilization regime required by the grasses.

Step 2: Redefining the problem
The RP reanalyzed the problems that became evident after the irrigation system was turned off and concluded:

- The rate of soil formation using just grasses and shallow irrigation was both too slow and too

shallow to overcome the deleterious properties of the raw tailings.
- A range of vegetation life-forms was needed that would persist on the TSF in the absence of irrigation and fertilization.

Steps 3 and 4: Redesigning solutions and applying technologies
To rehabilitate the flat top of the TSF, the RP selected a different design to apply:

- To create runoff and run-on areas that retained rainwater and limited any overland flows, the RP had engineers reform the surface into a system of internally draining minicatchments by reshaping dry tailings into shallow troughs surrounded by low mounds.
- To establish groundcover vegetation, the RP sowed grass seeds into the troughs and onto the

mounds. To increase the likelihood that these seeds would successfully establish and grow on the TSF, the RP collected seeds already growing on the TSF. (See Weiersbye et al. 2006; Weiersbye 2007.)

- To establish scattered trees on mounds, the RP planted tree seedlings grown in containers from locally sourced seeds.
- To kick-start biological processes to promote vegetation and soil formation, the RP covered mounds with mulch (garden refuse comprising shrub prunings, leaf litter, and a little soil). These piles of prunings also provided substantial wind resistance and captured soil particles carried in dust.

If successful, this pseudosavanna landscape would function like natural landscapes in the region, which have runoff and run-on zones at the 5 to 10 m scale.

To rehabilitate TSF walls, the RP had engineers reform these walls to have an angle of about 18° from the original 35°, which is close to the angle of repose for tailings. On the more gentle slopes the RP could simply till and sow seeds by machine.

To fix places along the base of TSF walls where leakage was evident, such as where the TSF was built over a former creek, the RP explored ways to use vegetation located in lands adjoining the base of the TSF walls to capture any plumes of seepage and reduce the chance of off-site contamination. The RP chose a mixed-species plantation-style design for use in these locations because research from around the world has demonstrated that no one tree species excels at all landscape processes. (See Weiersbye 2007; Weiersbye and Witkowski 2007.) Plantation tree species were specifically selected to transpire soil water and take up heavy metals.

Step 5: Continue monitoring and reassessing trends

The RP promptly began monitored the newly rehabilitated TSF tops and walls using the same restoration indicators as before, which reflected early trends in landscape processes such as the movement of tailings by wind and the retention of water, but added new indicators to assess tree establishment:

- To indicate the emerging role of vegetation to protect TSF surfaces from erosion, the RP measured the cover and size of planted trees and sown grasses along transects.
- To assess erosion on the rehabilitated TSF after each major wind and rainstorm event, the RP looked for signs of sediment transport such as plant hummocks and bare soil patches. If found, the RP located and mapped the potential source of the sediment.
- To evaluate the development of soil from raw tailings, the RP observed, for example, whether litter from vegetation was being decomposed and incorporated into the surface by fungi and soil invertebrates such as earthworms and termites.
- To assess whether contaminants, especially toxic heavy metals, were leaking from the base of TSF walls and beyond the plantations, the RP extensively collected water samples and tested water quality.

After monitoring these and other indicators over enough time to have reasonable confidence in the data, the RP evaluated landscape restoration trends.

For the tops of the rehabilitated TSF, the RP found that after only five years a dense and fine-scale patchy cover of grasses with a few trees had developed, which resembled the desired pseudosavanna (figure 7.9). To kick-start biological processes, the RP found a rich diversity of plant forms on mounds where garden refuse had been dumped (figure 7.10).

Based on monitoring for signs of wind erosion, the RP also found that the savanna-like landscapes were functioning to effectively control dust from blowing off the top of the TSF, which confirmed local empirical dust-storm observations. These pseudosavannas were also functioning to prevent water from flowing off the top of the TSF, which would create gullies in the TSF walls.

FIGURE 7.9. An open pseudosavanna landscape where raw tailings were used to form low banks and shallow depressions on top of a TSF, thus replicating the spatial heterogeneity found in natural landscapes. Variations in vegetation density reflect the bank and depression structure.

FIGURE 7.10. An example of vegetation development on the top of a rehabilitated TSF after more than five years where paddock-dumped garden refuse was the principle treatment; this refuse provided both substrate and microorganisms to accelerate soil-formation processes.

FIGURE 7.11. A profile of a soil forming under mulch applied to a mound on the top of a rehabilitated TSF.

The RP examined soil development and found that plant litter was being processed and buried by soil invertebrates in the litter. The activity of roots and other soil and biological processes (e.g., soil respiration) clearly indicated that soils were actively forming on the tops of rehabilitated tailing ponds (figure 7.11).

The RP monitored the TSF walls and found that after only five years (with no irrigation) a reasonable cover of grasses had established with only one initial application of fertilizer and lime. The RP monitored soil-surface condition indicators along transects on the rehabilitated TSF walls and found that soil-surface stability was providing a useful level of protection against wind and water erosion (figure 7.12), and that infiltration and nutrient-cycling potentials were only slightly below those expected from reference sites.

Out from the base of TSF walls, the RP found that plantations of mixed trees had successfully established and grown rapidly (plate 7.3). Tree growth was especially rapid along former drainage lines. (See Dye et al. 2008.) Collections of water from below these mixed plantations were not contaminated, which suggested that the selected tree species were actively transpiring soil and deeper drainage water, and taking up and immobilizing within plant tissues any heavy metals in this water.

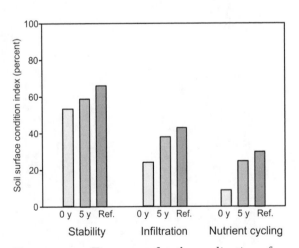

FIGURE 7.12. Five years after the application of new restoration technologies on TSF walls, values for soil-surface stability, infiltration, and nutrient-cycling indices had increased and approached those of reference sites.

Although after five years, restoration indicators suggested that trends were OK, it was too early for the RP to be certain that the new technologies put into practice to correct earlier problems had resulted in a self-sustaining pseudosavanna landscape. These technologies included inoculating raw tailings with real soil, creating internally draining basins on plateau tops, and planting vegetation adapted to local environments. Therefore, the RP continued to monitor indicators to confirm that these technologies, which were adaptively put in place to improve the rehabilitation of the TSF, were continuing to work.

Further Thoughts

Our description of restoring landscapes on TSFs in this chapter certainly does not imply that all problems have been solved. Satisfactory restoration of TSFs is still subject to ongoing investigations and new technologies. But, landscape restoration technologies similar to those described in this chapter are now being applied on mine sites around the world. And, if the procedures and principles described in this book prove successful in providing the framework for addressing the physical, chemical, and biological shortcomings of the tailings retained in TSFs, they could become industry standards for restoring mine sites.

In this chapter we did not cover the restoration of all types of mine-waste storage facilities. For example, we did not cover how RPs face the restoration challenges posed by coal ash storage facilities or the hazards of storing radioactive wastes. However, we are confident that RPs can successfully restore landscapes on all kinds of waste storage facilities by putting into practice our five-step adaptive procedure and its principles.

Chapter 8

Restoring Landscapes after Open-Cut Coal Mining

In this chapter we discuss the challenges faced by restoration practitioners (RPs) aiming to rehabilitate lands disturbed by open-cut coal mining operations. Large deposits of coal occur around the globe, notably in Australia, China, North America, and Russia. Coal deposits formed during Carboniferous, Permian, Triassic, and Jurassic geologic periods. Coals vary from soft brown (lignite), to soft black (bituminous), to hard black (anthracite). In many areas coal deposits lie exposed, or near the earth's surface, where they can be efficiently and economically extracted by open-cut (often referred to as open-cast) mining operations (figure 8.1). To reach coal deposits near the earth's surface, open-cut mining operations remove over-burden (regolith) materials to create large and extensive piles of waste material called spoil (figure 8.2).

Deeper coal deposits are mined by underground long-wall operations to create tunnels or galleries where coal is removed. Although underground mining may cause changes in surface hydrology due to collapse of galleries, tunnel operations produce very little waste or spoil. In this scenario we focus on the restoration of landscapes disturbed by open-cut coal mining operations because their large spoil heaps typ-

FIGURE 8.1. An example of an open-cut coal mining operation where massive drag-lines are used to strip away soil and over-burden and expose the coal layer.

FIGURE 8.2. A typical open-cut coal mining operation that has created extensive piles of spoil (over-burden). Note the erosion on the steep slopes of these spoil piles. The road on the top indicates that these spoil piles are up to 50 m high.

ically affect ten times more land than that affected by underground coal mining.

Restoring landscapes from large spoil heaps presents RPs with a number of challenges because raw (untreated) spoil materials are highly dispersive (unstable) and usually sodic (sodic saline or sodic alkaline; see glossary). These spoil characteristics are globally common on open-cut coal mines and worldwide cause RPs great difficulties when they try to establish vegetation on spoil. Here, our aim is to describe ways for RPs to successfully restore coal spoil heaps by applying our five-step adaptive landscape restoration procedure and by putting into practice the principles that underlie this procedure.

Setting the Scene

This scenario addresses restoring landscapes covered with large piles of coal-mine spoil. (See figure 8.2.)

Our scenario is based on our studies on restoring coal mines in the Bowen Basin of central Queensland, Australia (e.g., Spain et al. 1995), and on studies by colleagues around the world (e.g., Giurgevich 1999; Rethman et al. 1999).

The setting is a region where the climate is cool temperate with most of the annual precipitation of about 600 mm occurring during summer storms, but winter rains and snows also significantly contribute to total precipitation. Summer temperatures can be hot and winter temperatures freezing. The regional land use is predominantly cattle ranching (figure 8.3), and native vegetation in the region is woodland, which has been extensively cleared to form open pastures for cattle (figure 8.4). The soils are shallow lithosols (less than 20 cm deep) and generally have low nutrient status; because of this the country is used for ranching not farming.

Step 1: Setting goals

In this scenario, stakeholders required that sites being mined for coal be restored to landscapes that are suitable for grazing cattle (by reshaping spoil heaps to form gently rolling, grassy pastures); that yield little or no sediment into the lands surrounding the mine; and that have no off-site contamination of streams by pollutants (e.g., excessive salts because high-quality water is needed by cattle). Mine-site stakeholders con-

FIGURE 8.3. Cattle grazing a pasture created by clearing semiarid woodland.

FIGURE 8.4. A semiarid woodland (at right) that has been cleared (at left) to form an open pasture.

sidered these three goals adequate targets for the RP to achieve.

Step 2: Defining the problem

To achieve these goals, the RP critically analyzed a number of biophysical and socioeconomic problems created by the open-cut mining operations:

- Because of the way spoil materials were deposited by dragline operations (figure 8.1), spoil piles were unconsolidated mixes of materials (figure 8.2). This meant that reshaping these large, steep spoil piles into the landforms desired by stakeholders would be costly.
- Because of alkalinity (pH greater than 8) and sodicity (high concentrations of exchangeable sodium), spoil materials were highly dispersive and sides of piles readily eroded. (See figure 8.2.)
- Due to potential failures from pipe and tunnel erosion, dispersive spoil materials would not be suitable for building earthen banks to pond water for cattle and to form artificial wetlands.
- Because quality topsoil (pH around 7) was stripped off surfaces prior to mining and was stockpiled into hills, or buried, topsoil ended up being stored for so long that its biological activity was greatly reduced.

- Based on germination trials, the RP found that fresher stockpiles of topsoil contained high numbers of viable seeds, but unfortunately most were seeds from weeds—a legacy of former land use.
- Based on field and potting trials, the RP found that desirable pasture grasses readily grew in soils derived from regolith, especially when ameliorants such as gypsum and fertilizer were added. Rocky materials in the regolith rapidly weathered to a sandy loam texture.
- To evaluate trends in restoration indicators (see step 5), the RP analyzed the vegetation and soils on nearby natural woodland pastures (reference sites).

Steps 3 and 4: Designing solutions and applying technologies

After analyzing problems and evaluating findings from experimental trials, the RP designed a set of feasible solutions and technologies. In consultation with stakeholders, the RP selected and applied a number of technologies:

- To retain water within the rehabilitated landscape, the RP designed gently rolling pasturelands with a wetland. This landscape was built to be roughly saucer shaped, having widely spaced perimeter ridges (1,000 m apart), gentle slopes, and a central valley that drained into an artificial wetland (figure 8.5).
- To roughen pastureland surfaces to slow overland flows of water and reduce surface winds, the RP capped surfaces with topsoil and ripped slopes along contours.
- To establish pasture grasses and forbs on these rehabilitated surfaces, the RP sowed a seed mix of pasture species known to suit the prevailing climate and soil type, and known to be valued by local cattle producers; this seed mix included short-lived acacia shrub species.
- To ensure good shrub, grass, and forb growth after rains, the RP applied fertilizer.

FIGURE 8.5. A rehabilitated landscape shaped from coalmine spoil into gently rolling pastureland with low ridges and a valley draining into a wetland.

Step 5: Monitoring and assessing trends
After applying the technologies described above, the RP initiated a protocol to monitor indicators reflecting landscape processes on rehabilitated sites:

- To assess retention of water after storms, the RP looked for signs of on-site erosion (e.g., gullies and rills) and losses of sediments and other waterborne contaminants off-site.
- To evaluate the establishment of pastures, the RP annually measured the abundance and diversity of plants.
- To assess soil development in rehabilitated pastures, the RP annually collected soil samples and measured pH and available nitrogen and phosphorus for comparison with data from reference sites. To further assess soil development, the RP estimated soil-surface condition indices on the rehabilitated land-

scapes and on reference sites, especially with regard to soil slaking and dispersion indicators. (See chapter 14.) The RP required these data on indicators to evaluate trends as soon as possible so that, if necessary, adjustments to the design and technologies could be quickly applied.

After six years, the RP examined the monitoring data and found a number of positive trends toward those expected, but also some negative trends. (Note: For this scenario, data trends are generalized, but they are based on our studies on coal mines, particularly those located in the Bowen Basin, Queensland, Australia.)

- Pasture plants rapidly established on landscapes reshaped as gently sloping pasturelands. The RP found that grass densities on rehabilitated sites exceeded those found on nearby unmined pastures (figure 8.6a), but the diversity of all pasture plants remained well below that expected (figure 8.6b). This was probably because the seed mix was mostly exotic pasture grasses, and once established, they tended to exclude other plants in the mix such as forbs. The RP also found that native pasture plants were slow to establish, possibly because they were inhibited by the elevated concentrations of plant-available phosphorous applied as fertilizer to soils. In Australia, most native plants are adapted to soils low in available phosphorus.

- The fertility of the soils, as measured by available soil nitrogen and phosphorus on the rehabilitated pasturelands, exceeded levels found on unmined pastures in the area (figure 8.7). The RP expected this finding because of the application of fertilizer.

- The salinity of topsoil (0 to 5 cm) was initially fairly high (figure 8.8), but this diminished with time as leaching processes took effect. Although the salinity trend was toward that expected for natural pastures, the RP found that after six years soil sodicity (high exchangeable sodium) was still so high that soils remained dispersive (unstable).

- Due to the soil dispersion and slaking during rainfall, the RP found that after six years the riplines installed along slope contours had largely filled with sediments (figure 8.9).

- After only one year, the RP also found rills and gullies on rehabilitated slopes; these erosion features numbered as high as forty-five along a 100 m transect, and averaged 60 cm wide and 12 cm deep. Some gullies were over 1 m deep (figure 8.10). These erosion features imply

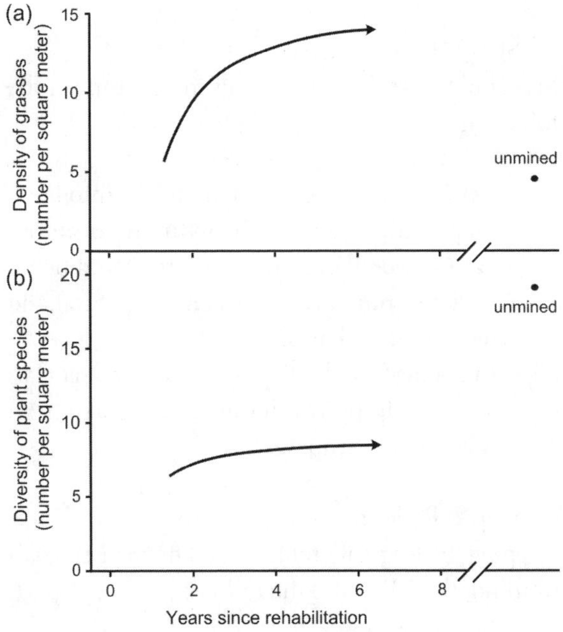

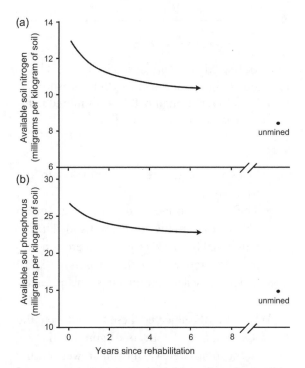

FIGURE 8.6. (a) Trends in the density of grass plants; (b) diversity of pasture species on rehabilitated pasture sites toward those expected from nearby unmined pastures.

FIGURE 8.7. (a) Mean available soil nitrogen; (b) mean available phosphorus in samples collected from the 0 to 5 cm depth on rehabilitated mine-spoil sites and on unmined pasture sites.

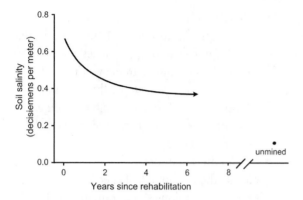

FIGURE 8.8. Soil salinity in the top 0 to 5 cm on rehabilitated mine-spoil sites compared with top-soil from unmined pasture sites.

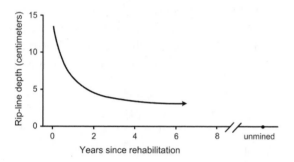

FIGURE 8.9. Mean depth of rip-lines on rehabilitated sites of increasing age. Depth is measured from bank tops to trough bottoms. Unmined natural pasture sites do not have rip-lines.

that soil dispersion remains a problem for the RP to solve.

- There was leakage of alkaline water (mean pH of 8.6) from the created wetland. The RP found that this leakage was from erosion tunnels formed within the dispersive spoil materials used to construct the wetland banks.

When the RP subjected these trend data to the *trends OK/trends not OK test* (see figure 1.1 in chapter 1), some trends were OK but others were clearly not OK. Perhaps the most serious failure indicated by monitoring was the severe gully erosion on the rehabilitated pasturelands. (See figure 8.10.) This made these landscapes unsuitable for grazing cattle.

Also, the leakage of saline waters from the created wetlands was contaminating the water the cattle would need for drinking.

The stakeholders conducted a cost-benefit analysis. They found that costs of restoration had been very high and that the potential returns from grazing enterprises on the rehabilitated pastureland was projected to be very low. They concluded that remedial actions were needed.

Trends Not OK: Follow the Adaptive Learning Loop

To decide what actions to take, mine-site stakeholders and the RP gathered in a workshop where they went through an adaptive learning loop process. (See figure 1.1.) In this workshop, they reevaluated their goals, reanalyzed their problems (seriously eroding and leaking landscapes) and designed new solutions.

Step 1: Resetting goals
Stakeholders revised their goals to accomplish the following:

- Reform and revegetate spoil heaps into landscapes that blended in with surrounding natural woodlands so that there would be no obvious boundaries between the natural and restored woodlands.
- Encourage the development of natural habitats for wildlife by restricting cattle grazing on restored landscapes.

Step 2: Redefining the problem
Keeping the original problems and the revised goals in mind, the RP found the following:

- Because less spoil material would be moved, land-forming operations to create natural-appearing woodlands would be less costly than forming gently rolling pasturelands and wetlands.

FIGURE 8.10. An example of a gully cutting into a slope reformed from spoil heaps created by open-cut coal mining. Grass plant root systems were unable to sufficiently ameliorate the dispersion properties of the spoil.

- By a thorough sampling of available spoils, relatively nondispersive materials could be selected and used as surface topsoil on rehabilitated sites. To improve topsoil dispersivity and slaking properties, the RP also found that gypsum could be applied to surfaces.
- To obtain a seed mix of species adapted to regional climates and soils, sufficient numbers of viable seeds could be collected from local woodland plants.

Steps 3 and 4: Redesigning solutions and applying technologies

Given the changed goals and findings from analyses, the RP designed, selected, and applied a number of new technologies:

- Small internally draining catchments were formed from spoil (figure 8.11) with a ratio of pond area to total catchment area of 1 to 1.5, which reduces loading of spoil with deep bodies of runoff water after storms. The RP tested infiltration rates using simulated rain and found that water infiltrated evenly across these small catchments.
- To improve surface soils for growing plants, catchments were capped with available topsoil and with relatively nondispersive spoil materials, which were also treated with gypsum to further lower dispersivity. Capping also positions the more dispersive spoil materials farther below the surface, where they are less likely to cause problems such as tunneling.
- To revegetate small catchments with species found in nearby natural woodlands, the RP sowed seeds of trees, shrubs, and grasses collected from local sources. Sowing rates were set to create plant densities similar to natural woodlands.

FIGURE 8.11. An example of a landscape reformed into small catchments (about 50 m from rim to rim) from spoil heaps produced by open-cut coal mining.

- To favor native plant species over exotic grasses and weeds, the RP did not apply fertilizers.

Step 5: Continue monitoring and reassessing trends

After applying these new technologies, the RP continued to monitor indicators previously measured, but also did the following:

- Examined internally draining catchments for any signs of leakage (e.g., tunnels through the catchment banks built from spoil).
- Measured the establishment and abundance of native trees and shrubs in addition to grasses.
- Estimated more carefully the soil-surface condition indicators, particularly in relation to the amount of litter and its state of decomposition, and the behavior of soil slaking. These indicators are especially important for estimating soil-surface stability and nutrient-cycling indices. (See chapter 14.) These indices reflect the complex processes involved in forming soil from spoil. (See Spain et al. 1995.)

After six years the RP evaluated trends in restoration indicators:

- There were no signs of leakage from the small internally draining catchments. Although the RP found some small rills on the sides of banks, all water and sediments were being contained within catchments.
- Native trees, shrubs, and perennial grasses readily established and grew within the small catchments (figures 8.12). The RP found that densities for trees and shrubs in catchments were toward those found on nearby unmined woodlands (figure 8.13).
- Trends in soil-surface condition indicators on rehabilitated spoil sites were also toward values found on unmined sites (figure 8.14). The RP concluded that these indicators provided evidence of soil formation processes (e.g., incorporation of organic matter from plant litter into the soil).

The RP then continued to monitor indicators for a few more years to confirm that restoration trends

FIGURE 8.12. After four years, native trees, shrubs, and grasses successfully established in small catchments created on rehabilitated coal-mine sites.

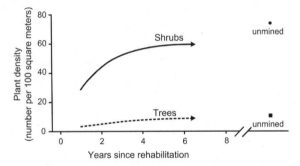

FIGURE 8.13. Density of trees and shrubs on rehabilitated sites shaped as small catchments and on nearby unmined woodland sites. Symbols: circle = shrubs; square = trees.

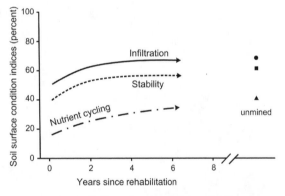

FIGURE 8.14. Soil-surface condition indicators on rehabilitated sites and on nearby unmined woodlands. Symbols: circle = infiltration index; square = stability index; triangle = nutrient-cycling index.

were OK. The RP presented these findings to stakeholders, who were satisfied that goals were being achieved and that restoring woodland landscapes from mine spoil was succeeding.

Further Thoughts

Because waste materials (spoils) associated with coal deposits almost always have adverse properties such as dispersivity and sodicity, RPs face significant challenges when using these materials to create stable landforms, to form new soils, and to revegetate these new landscapes. There are no shortcuts to reducing or isolating these adverse and variable spoil properties. The saying "know your enemy" applies here; in this case, RPs can characterize the properties of the materials emerging from the pit and, if feasible, sort spoils into heaps related to type and cost of amelioration treatments. For example, as over-burden is removed above coal deposits, any topsoil, subsoil, and other regolith materials with favorable properties can be separately stockpiled for later use to cover newly constructed landscapes. Over-burden with unfavorable properties, such as high sodicity and high pyritic (acid sulphate) concentrations, can be buried deep when creating new landforms.

Chapter 9

Restoring Rangelands with an Overabundance of Shrubs

This scenario describes the problem of restoring rangelands that were once open and grassy (figure 9.1) but have lost their groundcover and have become dense with unpalatable shrubs (figure 9.2). These dramatic changes are usually the result of many years of heavy grazing by livestock and, in numerous cases, by feral animals, such as rabbits, goats, camels, horses, and donkeys. Because of low groundcover, these disturbed rangelands are now eroding. They include former grasslands and grassy shrublands, such as the chenopod shrublands in southern Australia and the sagebrush shrublands in the western United States.

This problem of overly dense shrubs is common to rangelands around the world and is often described as *rangeland desertification*. (See, for

FIGURE 9.1. A highly functional rangeland with a high cover of palatable perennial grasses and a few scattered shrubs. There are few signs of accelerated soil erosion. Denser shrub and tree associations are confined to drainage lines in the background.

FIGURE 9.2. A monitoring transect in a dysfunctional rangeland where grasses have been replaced by dense unpalatable shrubs. Bare soil areas between the shrubs have high rates of wind and water erosion.

example, Reynolds and Stafford Smith 2002; Tongway and Ludwig 2002b.) For an easy-to-read account of the many problems caused by dense shrubs in rangelands, we refer readers to a delightful book by our colleague Jim Noble (1997).

Grassy rangelands that have become shrubby typically have signs of excessive soil erosion such as surface sheeting and rilling. *Sheet erosion* is the progressive removal of very thin layers of surface soil by water flowing broadly across extensive areas. (See plate 9.1.) There are few, if any, sharp discontinuities to demarcate these eroded areas. Sheet erosion typically occurs on gentle slopes (less than 2 percent) where soil surfaces are exposed. *Rill erosion* is the cutting of minor channels (less than 0.3 meters deep) by flowing water running down the steeper parts of a hillslope. (See plate 2.1.) The channels typically occur on slopes greater than 2 percent and are a sign that water is flowing rapidly and carrying materials such as soil particles, litter, and seeds with it. Gul-

lies are also channels cut by flowing water but are defined as being more than 0.3 meters deep. (See plate 2.2.)

In extreme cases of sheet erosion, the flows of water and wind can cause wholesale loss of soil A horizons and expose areas of hard-setting B horizons between remnant patches of vegetation (plate 9.2). These bare areas of soil are referred to as scalds or hardpans, which can be very extensive. We addressed the restoration of scalds in chapter 5.

A high density of shrubs also causes problems with rounding up livestock, because animals are hidden among shrubs in small disparate groups. This increases the time and financial cost of gathering stock for branding, culling, weaning, drenching, and other animal management activities.

In this chapter our aim is to illustrate for restoration practitioners (RPs) how to put principles into practice to restore shrubby rangelands. This scenario is based on our experiences, and those of our colleagues, with repairing damaged rangelands. (See, for example, Whisenant 1990; Friedel et al. 1996; Ludwig et al. 1997; Tongway and Ludwig 2002b; Milton et al. 2003; and Tongway et al. 2003.)

Setting the Scene

In this scenario we set the following conditions:

- The climate is semiarid (mean annual rainfall of 400 mm), seasonal (dry summers, wet winters), and temperate (continental, not coastal, so temperatures can range from minus −10°C to plus 45° C).
- The vegetation is an open grassy, low woodland (prior to becoming shrubby).
- The landscape is highly patterned or heterogeneous (i.e., banded vegetation). (See Tongway and Ludwig 2001, 2005.)
- The soils in shrubby areas are of low fertility, especially in available nitrogen and phosphorus (less than 25 and 10 mm per kilogram of soil, respectively). (See Tongway and Ludwig 1997.)

Step 1: Setting goals

We set this scenario in a rangeland region where increased shrub density has reduced the productive capacity and economic viability of pastoral enterprises. Stakeholders in these rangelands met in a workshop and established a number of restoration objectives:

- To reduce roundup times and costs, RPs must decrease the density of unpalatable shrubs.
- To reduce soil erosion, RPs must regenerate the cover of self-sustaining pasture grasses.
- To reduce total grazing pressures to promote grass recovery, RPs must remove or control feral herbivores such as goats and rabbits.

Also to reduce grazing pressures, the stakeholders set themselves a goal to strategically manage grazing by their sheep (e.g., avoid grazing of drought-stricken paddocks).

Stakeholders were confident that, if they and RPs could achieve these goals, their pastoral enterprises would be more profitable.

Step 2: Defining the problem

In the workshop, RPs and stakeholders examined the problem of unpalatable shrub increase and soil erosion:

- They analyzed causes of unpalatable shrub increase and concluded that it was due to overgrazing by domestic livestock (in this scenario, sheep), feral animals (e.g., rabbits and goats), and kangaroos. These animals grazed and browsed grasses and palatable shrubs down to a very low groundcover compared to that seen on reference sites. (Compare figure 9.1 with 9.2.)
- Because unpalatable shrubs are not browsed, and now have little or no competition from grasses and palatable shrubs, they increase in density.
- Because these shrubby landscapes have low cover at ground level, they permit overland flows to be excessive.

- Development of artificial watering points (bores, wells, and earthen tanks) within paddocks, which improved the availability of water for stock, and also for ferals, prolongs and intensifies grazing. (See our discussion on overgrazing and watering points in chapter 5.)
- Soil erosion was always the most severe near watering points, as might be expected, and damage declined with distance away from water (i.e., a grazing gradient). (See Landsberg et al. 2003.)
- Fire frequency analysis showed that, as a consequence of low grass biomass to provide fuel, fires no longer occurred. Historically, fires had been frequent, which controlled shrubs, especially at seedling and sapling stages.

Stakeholders and RPs also analyzed long-term rainfall records and found that rains were highly unreliable in terms of amounts and seasonality. Droughts of three to five years' duration were common. During major rainfall events they observed that rangelands with high grass cover retained (captured) overland flows of water at a fine scale of about 20 to 30 cm. (See figure 9.1, and also figure 2.1 in chapter 2.) Also, in rangelands with dense shrubs, patterns of runoff indicated that losses of water from the landscape were high because of very little obstruction by ground layer vegetation. (See plate 9.1.) Any water retained in the landscape primarily benefitted shrubs.

In shrubby rangelands, RPs and stakeholders also analyzed soil surfaces and soil properties and observed that soil surfaces were found to have developed smooth, hard crusts under raindrop impact; hence, runoff was high and infiltration in areas of bare soil was very low. Except for some soils along drainage lines, most of their soils did not have swell/shrink properties, that is, whether soils swell on wetting and shrink again on drying. This means that natural soil cracks do not form and surface crusts are nearly continuous. (See chapter 5 for a discussion of soil swell/shrink properties.) They also observed that soil surfaces in bare overgrazed areas tended to be unstable because they readily slaked or dispersed, whereas soils within vegetation patches were noticeably more stable. (See slake and dispersion tests in chapters 14 and 15.)

Stakeholders conducted economic analyses, which showed that roundup costs were very high due to the difficulty of finding and moving livestock through dense shrubs.

Steps 3 and 4: Designing solutions and applying technologies

To reduce shrub densities and allow grasses to reestablish, RPs and stakeholders designed a strategy that combined mechanical and chemical treatments:

- Based on a comparison of different mechanical treatments for killing shrubs, such as blade-plowing, chaining, pushing, stick-raking, and burning, they selected blade-plowing (figure 9.3). Experimental trials had demonstrated that blade-plowing efficiently deals with small to large shrubs by (1) severing the roots, (2) uprooting the crown, and (3) leaving most of the severed roots buried. These factors significantly increase the mortality of the shrubs in treated rangelands. (See Wiedemann and Kelly 2001.)
- To target any shrubs emerging after blade-plowing (from sprouts or seeds), they designed aerial applications of chemicals (figure 9.4). They only selected chemicals known from experimental trials to effectively kill unpalatable shrubs. (See Noble et al. 2001.)
- To promote the establishment of grasses, they designed a plan to sow grass seeds at a time when significant rains were expected (determined by tracking daily weather reports).
- To promote native perennial grasses, they chose to sow annual grasses as nurse plants.

FIGURE 9.3. A blade-plow used to control rangeland shrubs by severing their roots below ground.

Figure 9.4. Aerial application of chemicals for controlling rangeland shrubs. Photograph courtesy Jim Noble.

They found that it was difficult and expensive to obtain native perennial grass seeds. Their aim in using annual grasses was to stimulate the germination of seeds of native perennials already present in the soil.

Step 5: Monitoring and assessing trends
Before and immediately after sequentially applying the physical, chemical, and biological technologies selected above, RPs also selected a number of restoration indicators to monitor. They specifically chose to assess indicators reflecting changes in landscape functionality and soil-surface condition on treated areas and relative to reference sites, which included the following:

- Shrub and grass cover
- Plant species composition
- Weed species establishment
- Signs of soil-surface sheeting (pedestal erosion around plants)

- Signs of distinctive runoff pathways (rills)
- Indicators of soil-surface stability and infiltration and nutrient-cycling potentials

After three years, when RPs examined trends for progress toward restoration goals on sites where mechanical and chemical treatments had been applied, it was obvious that a number of serious problems remained:

- A reduction in overgrazing had not been achieved because of high numbers of feral animals, which were consuming emerging grass seedlings.
- A reduction in runoff and erosion was not achieved because grasses failed to establish and improve groundcover.
- Shrub density was not reduced but had actually increased. Shrubs had resprouted and reseeded in root-plowed areas because of soil disturbance. The single application of a chem-

ical treatment had helped kill newly emerging shrubs, but the chemicals were too expensive to repeatedly treat sprouts and seedlings.

After examining their current goals and the likely costs of reducing feral animals and applying additional chemical treatments, stakeholders concluded that continuing was clearly uneconomic. Because their current restoration goals could not be achieved, their rangelands would have to be abandoned and placed on the market for sale. Land use would probably change from livestock grazing to a different type of enterprise, such as game hunting, as had already happened on other shrubby country in other regions.

Although problems remained and trends in many indicators were not OK, there were some positive trends, so instead of giving up, RPs encouraged stakeholders to work through the adaptive learning loop. (See figure 1.1 in chapter 1.)

Trends Not OK: Follow the Adaptive Learning Loop

In a workshop setting, stakeholders first reexamined their goals.

Step 1: Resetting goals
They agreed that their original goals to reduce shrub densities, soil erosion, and grazing pressures were still appropriate. Then they analyzed in depth the problems noted earlier.

Step 2: Redefining the problem
These analyses led the practitioner and stakeholders to the following conclusions:

- To reduce grazing pressure and encourage sown pasture grasses, they must more effectively reduce feral animals.
- To reduce soil erosion after rains, they must find ways to slow and retain flows of water (runoff) within the landscape.

- To reduce shrub densities, they must more effectively apply mechanical and chemical treatments.

They deemed particularly important the retention of water in the landscape to enhance grass establishment and growth, because a good cover of grasses would then provide a positive feedback so that more water is captured during future rainfall events. (See figure 2.11 in chapter 2.)

Steps 3 and 4: Redesigning solutions and applying technologies
Based on these analyses and conclusions, stakeholders and RPs agreed to apply the following designs and treatments:

- To reduce total grazing pressure, which would promote native perennial grasses, they worked out strategies for (1) moving domestic stock out of paddocks being treated; (2) capturing and selling feral animals (e.g., goats) from treatment paddocks on a regular basis; (3) turning off watering points within treated paddocks to discourage use by feral animals; (4) refraining from restocking treated paddocks until grasses had fully recovered; and (5) after recovery, strategically resting the paddock from grazing (e.g., spelling after rains) to allow plants to grow more substantial root systems.
- To reduce shrub densities and sow annual nurse grasses at the same time, they attached a seeder to a blade-plow.
- To redistribute water slowly across the landscape (water spreading), especially in areas showing signs of rill and gully erosion, they built contour banks and troughs (figure 9.5). Their design used an upslope bank (by pushing earth from downslope) to arrest overland flows and distribute water through small openings into shallow downslope troughs, which, when full, overflow to gently release water across the entire slope contour. They precisely positioned earthen banks along contours

Cross section

Horizontal section

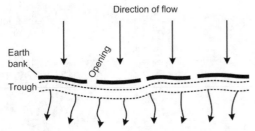

FIGURE 9.5. A design sketch of how earth banks were constructed to slow and spread surface flows of water.

FIGURE 9.6. Brush packs constructed along a contour line.

using laser technology. Their overall aim of slowing and spreading water across the entire hillslope was to enhance infiltration and the establishment and growth of grasses over the landscape. (See chapter 5 to see how larger water-ponding banks were used to repair eroding rangelands.)

- To reduce erosion in areas showing severe surface sheeting, they positioned brush packs (piles of twigs and branches) along contours (figure 9.6). Brush packs are known to function as "leaky" weirs by slowing surface-water flows, enhancing infiltration, and trapping sediments, litter, and seeds. (See Tongway and Ludwig 1996.)
- To control any unpalatable shrubs emerging in spite of these treatments, they spot-sprayed chemicals.

Step 5: Continue monitoring and reassessing trends

To evaluate restoration trends following the application of their new designs and technologies (e.g., water-spreading banks, brush packs) on treated sites,

RPs monitored the previously measured indicators. After measuring indicators for six more years on water-spreading areas, they found the following:

- Earthen banks and troughs were effectively slowing and spreading surface flows and promoting the establishment, growth, and reproduction of pasture grasses and forbs, especially in areas above earth banks (figure 9.7).
- Soil-surface condition indicators in water-spreading areas had greatly improved, including surface stability, infiltration capacity, and nutrient-cycling potential, and they were trending toward values expected from reference sites (figure 9.8). (Although generalized for the purposes of this scenario, these findings are based on one of us (DT) working

FIGURE 9.7. An example of the high grass cover found in rehabilitated rangelands where earth banks were used to spread water over the landscape.

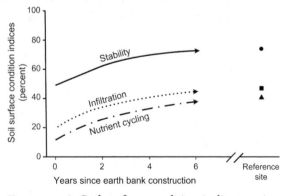

FIGURE 9.8. Soil-surface condition indicators six years after construction of water-spreading earth banks compared to values for reference areas. Symbols: circle = stability index; square = infiltration index; triangle = nutrient-cycling index.

with a stakeholder using earth banks to rehabilitate rangelands.)

For this scenario, brush-pack findings are generalized from two research papers. (See Ludwig and Tongway 1996; Tongway and Ludwig 1996.) From monitoring indicators on areas treated with brush packs, the RPs found the following:

- Brush packs had enhanced the establishment of native pasture species (figure 9.9), so that after ten years, recovery of palatable perennial grasses and palatable shrubs was clearly trending toward values expected from reference sites (figure 9.10).

FIGURE 9.9. An abundance of palatable perennial grasses, forbs, and subshrubs have established within brush packs. The brush packs prevented grasses being grazed down to ground level by sheep and kangaroos. Plants outside brush packs are scattered and mostly unpalatable ephemerals.

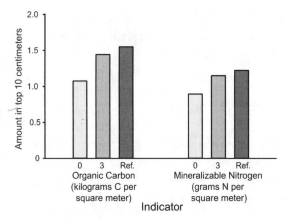

FIGURE 9.11. Values for soil organic carbon and mineralizable nitrogen within brush packs after three years compared to preconstruction (0) values and those for reference (Ref.) sites.

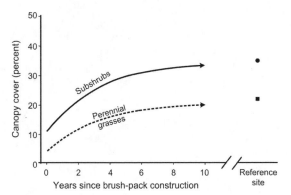

FIGURE 9.10. Canopy cover for palatable perennial grasses and subshrubs within brush packs after ten years was toward that expected from reference site data. Symbols: circle = subshrubs; square = perennial grasses.

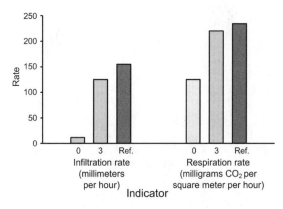

FIGURE 9.12. Rates for water infiltration and soil respiration within brush packs after three years since construction (0) compared to values measured on reference (Ref.) sites.

- Soil fertility within brush packs had increased. After only three years, they recorded significant increases in soil carbon and nitrogen concentrations in the top 10 cm of soil (figure 9.11).
- Soil infiltration and respiration rates had also markedly improved within brush packs (figure 9.12).

Taken together, these indicators show that soils within water-spreading and brush-pack treated areas

are becoming much healthier and better able to support perennial pasture grasses.

Restoration practitioners continued to monitor restoration indicators across the treatment sites until trend analyses confirmed that all trends were OK. For example, in time, an abundance of pasture grasses and litter was providing sufficient cover to ensure that runoff remained slow and diffuse, which greatly reduced surface erosion. Because shrub densities were now low, the ease of rounding up animals was greatly enhanced, which reduced costs. The pasture was periodically completely rested from grazing, resulting in more persistent plant growth and more sustainable ecosystems.

Further Thoughts

Although this scenario is hypothetical, from our experiences we have observed that ranchers who successfully restore their shrubby rangelands gain a notable improvement in their quality of life:

- An excessive amount of time is no longer needed to gather animals for market.
- Cash flow from sheep and feral goat sales is a significant ongoing financial benefit.

- Soils, plants, and animals are clearly healthier and more productive.

We have also observed that to successfully restore shrubby rangelands in the long term requires diligent ongoing treatments. For example, as unpalatable shrubs occasionally emerge (despite competition from other pasture plants, which function to keep shrub numbers low), they need to be spot treated with herbicides, or mechanically removed, before becoming old enough to produce seeds. Some herbaceous weeds also emerge in earlier stages of rehabilitation, and, if aggressive, they may need to be spot treated. With time, almost all weeds are excluded by vigorous native pasture herbs and grasses.

PLATE 2.1. Damaged landscape with active runoff and surface erosion processes. Note the "milky" runoff indicating how fine soil sediments are being removed from this landscape.

PLATE 2.2. Landscape where runoff is cutting a gully. Photograph courtesy of Aaron Hawdon.

PLATE 2.3. Example of a structurally diverse and highly functional landscape where resources are retained and habitats are available for a diverse fauna.

PLATE 3.1. Forest landscape has been buried by waste materials (spoil) created by mining. Gullies are cutting into the steep slopes, but this has not deterred someone from building a house on top of this mine-spoil heap.

PLATE 3.2. A highly functional landscape having a dense cover of shrubland vegetation to protect soil surfaces from wind and water erosion.

PLATE 3.3. A heavily grazed dysfunctional landscape having very little shrubland vegetation to protect soil surfaces from erosion; this landscape is located near the one shown in plate 3.2.

PLATE 4.1. Typical soil aggregate found in topsoils of open forests at Gove.

PLATE 4.2. Landscape being prepared for mining bauxite by pushing open forest vegetation into windrows that are burnt when dry. Photograph courtesy of Sue Gould.

PLATE 4.3. After mining bauxite (background) an ironstone surface remains (foreground).

PLATE 4.4. One-year-old revegetated site at Gove.

PLATE 4.5. Seven-year-old revegetated site at Gove.

PLATE 4.6. Twenty-year-old revegetated site at Gove.

PLATE 5.1. Woodgreen ponding bank 5: photo point in 1985. Photograph courtesy of Gary Bastin.

PLATE 5.2. Woodgreen photo point (seen in plate 5.1) in 1990. Photograph courtesy of Gary Bastin.

PLATE 5.3. Water ponding on an extensively scalded area. Photograph courtesy of Ray Thompson.

PLATE 6.1. An angle of repose waste-rock dump with active erosion compared to the natural landscape (foreground) where soil erosion can scarcely be detected.

PLATE 6.2. Dump of mixed regolith materials produced by open-cut mining.

Plate 6.3. Restoration practitioner making additional measurements that will be used to determine the effectiveness of strips of mulch in trapping sediments eroding from upslope.

Plate 7.1. The top of a tailings storage facility being covered with topsoil after capping with coarser materials. Photograph courtesy of Harley Lacy.

PLATE 7.2. Wall of a tailings storage facility with distinctive bands of different tailing materials.

PLATE 7.3. Plantations located at the base of tailings storage facilities.

PLATE 9.1. Sheet erosion in action. A broad shallow body of water is removing thin layers of soil in a rangeland.

PLATE 9.2. Scald erosion in a rangeland. Over 100 mm of soil has been lost by wind and water action, exposing a dispersive clay subsoil.

PLATE 11.1. Crimson Rosella, which utilizes complex habitats in the woodlands of eastern Australia, where its diet consists of buds, floral nectar, seeds, fruit, and insects. Photograph courtesy of Liz Harman.

Chapter 10

Renewing Pastureland Functions Using Tree Belts

In this scenario we describe the procedures and principles that restoration practitioners (RPs) can put into practice to improve a number of functions and values (e.g., retaining water and improving biodiversity) on farms where woodlands have been cleared for use as pasturelands. To create pastures and grow crops in many regions around the world, woodlands have been extensively cleared. This clearing of woodland trees alters the way landscapes function. (See Ludwig and Tongway 2000, 2002.) Tree clearing can cause runoff and loss of soil even in relatively gently sloping landscapes (figure 10.1). Attempts to avoid such problems by growing pastures for livestock, rather than farming crops, have often failed because closely grazed pastures provide very little resistance to overland flows, especially when rainfall events are intense. Even moderate rainstorm events can wash soil sediments, animal dung, and plant litter from hillslopes into riparian zones and can create further difficulties, such as the clogging of natural drainages.

Globally, societies are becoming aware of the

FIGURE 10.1. An example of runoff eroding soil from a recently cleared woodland slope.

importance of preventing soil erosion and main-taining biodiversity in open pasturelands created by clearing woodland and forest trees. This aware-ness has resulted in a number of responses from different community groups (stakeholders). For example, Australian farmers have taken positive action by organizing themselves into *Landcare* groups. (See www.daff.gov.au/natural-resources/landcare/ [accessed 20 February 2010].) These groups aim to improve landscapes by planting trees in belts aimed at preventing severe runoff and ero-sion events, avoiding further declines in species diversity by increasing habitat, and providing shel-ter for livestock.

We construct this scenario on the use of trees planted in belts to retain runoff, because tree belts illustrate the *leaky* weir or dam principle in which flows of water are slowed and deenergized but not ponded. This principle is also illustrated by brush packs aligned in rows along contours. (See chapter 9.) As with brush packs, tree belts avoid the need to construct earthen banks to slow and spread overland flows (runoff). (See chapter 5.) Instead of human-made earth barriers, tree belts provide living, self-sustaining flow regulators. Typically, water-spreading and water-ponding banks are applied on slopes of less than 2 percent, and tree belts are deployed on slopes greater than 2 percent. Tree belts aim to mimic the structures and functions of vegetation patches found in nat-ural woodland landscapes. (See Tongway and Ludwig 2005.)

Although hypothetical, the following scenario is based on our experiences and those of others with restoring pasturelands created by clearing trees in savannas, woodlands, and forests. (See, for example, Whisenant 1990; Ludwig and Tongway 2002; Ellis et al. 2006; Munro et al. 2007; and Ryan et al. 2010.)

Setting the Scene

In this scenario, we set the pasturelands that RPs are restoring in a region having the following attributes:

- A temperate climate, which is character-ized by cool, rainy winters and springs, and warm, dry summers and autumns. Except during drought periods, winter rains are relatively reliable, whereas summer rains occur as infrequent and scattered storms.
- Woodland vegetation, which occurs as highly fragmented, isolated remnant patches because of clearing. Grasslands are con-fined to exposed, drier ridges and forests to wetter zones along drainage lines, such as creeks and rivers.
- No coarse woody debris (logs, branches) on surfaces, because woody debris was removed during tree clearing to create open pastures for sheep rather than for cattle.
- Wild herbivores, such as rabbits, occur in the pastureland, but their grazing impacts are considered minor.

Step 1: Setting goals

Members of a Landcare group met and agreed to have RPs plant trees in belts within their pas-turelands to provide more woodland habitat for a range of wildlife species, create wildlife corri-dors by linking existing woodlands, provide more shade and shelter for their sheep, and reduce the flashiness of runoff and erosion events on their pastures. Reducing runoff from pastures has the benefit of reducing flow rates into drainage lines such as streams, which reduces, for example, severe erosion along stream banks (fig-ure 10.2).

Step 2: Defining the problem

Landcare members and RPs held a series of work-shops, which included researchers, to examine both the positive and negative aspects of using belts of trees in their pasturelands.

Positive factors
Planting belts of trees along contours in their pastures would have a number of benefits:

FIGURE 10.2. Excessive overland flows from upslope pastures have caused stream bank erosion, which has exposed the roots of a woodland tree.

- To keep the original pasture soils intact, RPs could plant trees across slopes in ways to minimize soil disturbance
- To provide some protection of existing pasture surfaces, RPs could also plant trees in belts to keep most of the pasture vegetation (grasses) intact.
- To provide corridors for the movement of wildlife between woodland islands, which were currently distant from one another, RPs could plant belts of trees to connect across adjoining pastures and farms.
- To maximize the survival of trees, RPs could select species adapted to the local climate and available as nursery-grown stock.
- To reduce environmental stress on their sheep, RPs would plant trees that provided animals with shade on hot days and shelter on cold windy days.
- To keep land preparation and planting costs low and affordable, farmers could cooperate to provide RPs with the machinery and labor needed to plant belts of trees.
- To improve soil properties, RPs could plant trees that produced litter and woody debris, promoting soil invertebrates such as earthworms, which function to develop the soil porosity needed to enhance infiltration rates and facilitate gas exchange (soil respiration). The contribution of these organisms is a largely forgotten component of biodiversity.

Negative factors

Analysis of the situation by Landcare members, RPs, and researchers also exposed some difficult challenges:

- Due to heavy trampling and grazing by stock over many years, they found that pasture soils had become smoothed and compacted so that pasture surfaces lacked the roughness needed to reduce high overland flow rates during storm events.
- Because, over the long term, a lot of the grass production consumed by sheep was exported out of the local landscape, they found that pasture soils were low in organic carbon.
- Because pasture grasses have shorter life cycles, they found poor nutrient cycling compared to that found in natural woodlands, which consistently produces woody litter that decomposes slowly over decades.
- Because of a combination of trampling and raindrop action, they found that pasture soil surfaces had hard physical crusts. (See *physical soil crusts* in glossary.) These hard crusts have low water-infiltration rates causing runoff to commence earlier during rainfall events so that less rainwater is stored in deeper rooting zones and landscapes are drier than normal.
- Because of the time it takes for trees to establish and grow, the RPs knew it would take some years before tree belts would significantly reduce overland flow rates.

Steps 3 and 4: Designing solutions and applying technologies

To meet the challenges revealed by the above analyses, Landcare members, RPs, and researchers worked together to design ways to quickly and successfully establish tree belts across pastures.

Design factors

They took a number of factors into account when designing tree belts:

- Because the capacity of landscapes to trap and retain overland flow is a crucial restoration principle, and because sheep tend to follow fence lines, they knew that simply planting tree belts along property boundaries, which often have sheep trails running downslope, would fail to achieve restoration goals. Therefore, they selected a design where tree belts were precisely aligned along slope contours. This would maximize the interception and capture of overland flow while helping to reduce the impacts of sheep trails and camps along boundaries between tree belts and open pastures.
- To provide adequate control of overland flows, they used computer simulation models to estimate the optimal downslope depth or width of the tree belt needed to retain most of the water flowing in major rainfall events. Using long-term climatic data (e.g., rainfall amounts and intensities), models predicted that on gentle slopes 10 m wide tree belts were most efficient at capturing runoff, but on steeper slopes belts up to 20 m wide were predicted to be more efficient.
- On steeper slopes, they considered an alternative to 20 m wide tree belts, which was to use more 10 m wide tree belts within the pasture. They chose this alternate design in cases where their machinery was set up for preparing ground and planting trees in 10 m wide belts.
- To ensure that the tree species planted would be adapted to the regional climate, they sourced species from local natural woodlands. Tree species selection also took into account predicted trends due to climate change (e.g., drier winters and springs) and that pasturelands tend to be open, drier, and windier environments than natural woodlands.
- To accelerate initial resource capture, they chose to plant some shorter-lived trees and shrubs, but in the long term their tree-belt management plans used long-lived trees.

- To reduce the time taken for tree belts to effectively regulate overland flows, they planted some tree species known to produce litter in a shorter time than others.
- To reduce through-the-fence tree browsing pressures, especially when gates were closed to exclude grazing from inside tree belts, they planted less palatable (e.g., thorny) shrub species next to fences.

Application factors

When planting trees in belts, they considered a number of other factors:

- So that tree planting could be done in a single efficient operation, they prepared (sprayed and tilled) all areas to be planted in advance.
- To protect plantings during the establishment phase, they fenced tree belts to control sheep and kangaroo browsing.
- To facilitate strategically timed grazing of tree belts by sheep at later stages, they installed gates in fences. They knew that controlling the level of grazing pressures is very important, because excessive animal disturbance within tree plantations slows or reduces the buildup of litter and woody debris.
- To prevent competition between weeds and tree seedlings, they spot sprayed weeds with herbicides.

Step 5: Monitoring and assessing trends

Members of the Landcare group and RPs decided to monitor tree belts in two phases: (1) early assessments of the establishment of the planted trees, and (2) later observations on the extent to which tree belts were regulating overland flows.

They found that early monitoring allowed them to quickly replace any trees that died. This avoided any gaps in tree belts where overland flows might break through. Filling gaps is an aesthetic consideration too. Early monitoring also allowed them to quickly repair fences.

Later, they conducted monitoring of resource reg-

ulation after each significant rainfall event by examining fence lines on the upslope and downslope sides of tree belts:

- On the upslope side, they observed how much litter and dung was accumulating in the first few meters above the first line of trees.
- On the downslope side, they looked for any evidence of outflows through the tree belt, such as deposits of litter and soil sediment below the last line of trees. They also noted any signs of soil erosion, such as rills extending out from the lower side of the tree belt.

Although this monitoring was done on a relatively casual basis, the basic rules of monitoring were not broken because *observations were focused on landscape processes and were spatially referenced so that rapid repair actions could be taken.*

After the tree belts were well developed, they monitored the goods and services being provided by tree belts, which included observations on the following:

- The abundance and diversity of native fauna using tree belts such as woodland birds.
- Use of tree belts by sheep as shade and shelter.
- The amount of litter and woody debris accumulating within tree belts. In early years, soft leaf litter typically dominates, but as trees mature, coarser woody debris becomes dominant.
- The extent to which litter decomposition activities (chewing by invertebrates, consumption by fungi and bacteria) were developing dark humus layers in the soil.

After fifteen years, RPs and Landcare members assembled and examined their monitoring observations and data. They found the following:

- Planted trees had successfully established and grown to form thick, robust belts several meters high within their pasturelands (figure 10.3).

FIGURE 10.3. An example of a 15-year-old tree belt viewed from upslope looking down to the belt. The open pasture (foreground), although not overgrazed, has a low capacity for slowing overland water flows.

- Tree belts appeared to be effectively capturing runoff from the pasture above the belt and from the zone of bare, compacted, and crusted soil immediately above the belt where sheep formed trails (figure 10.4).
- A variety of birds, such as the rainbow lorikeet (figure 10.5), were now readily observed in tree belts.

In comparing their monitoring data, Landcare members also identified additional benefits of tree belts:

- Their sheep were benefitting from the shelter and shade provided by the tree belts. When gates were open, their sheep often took shelter and grazed within the tree belts, and when gates were closed, to reduce grazing pressures, their sheep took shelter and camped along the leeward boundary of the open pasture and tree belts.
- They found that a considerable amount of litter and woody debris was accumulating within the tree belts, and there were signs that these materials were being incorporated into the soil. This meant that the capacity of tree-belt soils to recycle nutrients was improving.
- After significant rainstorms, they found very few signs of erosion, such as rills or any litter and sediment being deposited downslope of the tree belts. This meant that high surface roughness and infiltration rates within tree belts were preventing runoff from penetrating through belts.

The Landcare group and RPs worked with researchers to verify their monitoring observations and records. The researchers used a rainfall simu-

Figure 10.4. A 15-year-old tree belt viewed along the slope contour showing how sheep traffic and camping close to the plantation has created a bare crusted soil with high runoff rates.

Figure 10.5. A woodland bird (rainbow lorikeet) commonly observed within established pastureland tree belts.

lator setup (figure 10.6) to apply the equivalent of a one-in-ten-year rainfall event (45 mm per hour) for thirty minutes to 90 m^2 tree-belt plots and to 210 m^2 pasture plots, which were located upslope of the tree belts. From this experiment they found a number of differences: (These findings are generalized from studies by Ellis et al., and Leguédois et al. 2008.)

- When the rainfall simulator delivered the equivalent of 24.4 mm of rainfall in thirty minutes, runoff was generated from the pasture plots after only about one minute, but runoff did not appear from the tree belt until after twelve minutes. These time-to-runoff differences relate to differences in water infiltration rates, which were 45.3 mm per hour in the tree belt plots, but only 31.2 mm per hour in the pasture plots.
- Runoff flowed faster and shallower down open pasture slopes than within tree belts (figure 10.7).
- There was virtually no resistance to overland flow in the bare hard-crusted zone immediately above tree belts where sheep trails were common (figure 10.8) and only slightly more resistance in the pasture, as indicated by twinkling light reflections in the induced turbulence. There was much greater surface resistance to flows within tree belts due to a layer of tree litter and debris (figure 10.9), which was up to 50 mm deep.
- Overland flows delivered a total of 19.8 kg of soil sediment off the pasture plots, but only 0.9

FIGURE 10.6. A rainfall simulator in action delivering water from sprinklers to pasture and tree-belt plots at about 45 mm per hour.

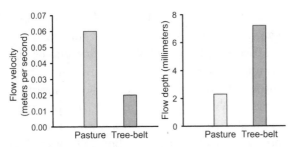

FIGURE 10.7. Runoff or overland flow velocities and depths produced by applying 24.4 mm of water in 30 minutes to pasture and tree-belt plots. See Ellis et al. (2006).

FIGURE 10.8. During simulated rainfall, an experimental plot located within the bare zone upslope of the tree belts provided little resistance to flows of water.

FIGURE 10.9. During rainfall simulation experiments, litter and woody debris within tree belts provided high surface resistance to flows of water and also filtered out soil particulates.

kg of this sediment came through tree-belt plots. This means that most of the sediment was trapped within the tree belt.

To predict whether adding tree belts (hypothetically) would reduce overland flows within and from small catchments, Landcare members, RPs, and researchers used a computer simulation model based on hydrological processes. (See Ryan et al. 2010.) The model predicted that 20 m wide belts of trees oriented along contours across steep hillslopes (up to 24 percent slope) within a creek catchment reduced average maximum overland flow velocities by 14 percent on

hillslopes with tree belts compared to those without tree belts.

Based on their observations and findings from monitoring, which were confirmed by experimental studies, Landcare members concluded that, after fifteen years, planting trees in belts in their pasturelands had achieved their restoration goals.

Further Thoughts

In early settlement times, very extensive tree clearing was regarded as essential for the success of farming enterprises: *Fewer trees means more grass.*

However, wholesale tree-clearing practices have greatly reduced habitats essential to the survival of native biota, and biodiversity conservation is now viewed as a major issue by society. Promotion in the popular press has kept conservation topics before the public. Such topics make good stories, for example, about how Landcare groups are planting patchworks of trees within farmlands to provide woodland corridors that allow threatened species of native fauna to travel between patches of remnant vegetation. Although we did not specifically focus on biodiversity conversation in our scenario here, restoring pasturelands to improve their capacity to retain water and soils by building belts of trees undoubtedly creates new habitats for a greater diversity of plants, animals, and microorganisms.

In designing tree belts, 100 percent resource capture need not be an objective, as long as belts of trees substantially deenergize overland flows. Also, maintaining slightly taller grasses within open pastures decreases the rate of overland flow through the pasture (farmers have observed this during rainstorms). To maintain taller grasses in pastures, farmers can strategically manage grazing pressures.

We also did not deal specifically with some of the more severe landscape degradation problems caused by tree clearing, such as those changes to landscape hydrology that have resulted in salinization of surface soils from rising saline water tables. In these landscapes, tree transpiration previously kept water and salts at depth because trees roots extract water from deep within soil profiles. Although fully satisfactory solutions to such complex and severe problems are very difficult to achieve, RPs can significantly improve landscape functions by applying the procedures and principles we described in this chapter.

Chapter 11

Restoration of Former Farmlands Near Urban Developments

In this chapter we describe how community groups are actively restoring former farmlands by putting into practice our principles and procedures of adaptive landscape restoration. In many regions around the globe, farmlands provide space within expanding urban areas for new housing developments and for green space (figure 11.1). To provide more functional green space near urban areas (e.g., more habitat for birds, walking trails, more native plants, and fewer exotic weeds), community groups are actively restoring former farmlands.

In recent decades around the globe there has been an upsurge in restoration activities by community groups, because as outer suburbs expand they take over more adjoining farmland fields and pastures, which were originally natural vegetation such as woodlands and forests. More and more natural habitats are being lost. The original landscapes may

FIGURE 11.1. Former farmland provides space for suburban housing developments.

have been totally cleared for farming or some remnant vegetation may be retained at the edge of housing developments (figure 11.2) and within farmlands (figure 11.3). All these disturbed landscapes present difficult challenges to community groups working on farmland restoration projects.

Our aim in this scenario is to document how community groups, working as restoration practitioners (RPs), can meet these challenges. We do this by presenting a landscape restoration scenario based on the experiences of one of us (DT) while working with community groups on projects in the Australian Capital Territory. We include this example as a scenario chapter in part 3 because these landscape restoration projects have only been active for a few years, whereas the case studies described in

chapters of part 2 are based on more than twenty years of restoration activity and monitoring.

Setting the Scene

The setting of this scenario includes former farmlands near the city of Canberra in the Australian Capital Territory (ACT), which is within New South Wales. (See figure 11.4.) The region comprises the following characteristics:

- Climatic conditions are temperate, continental, and mountainous. Summers are typically hot and dry, and winters, cold, foggy, and frosty. Summer temperatures average about 25°C in

FIGURE 11.2. Some remnant vegetation has been retained at the edge of a housing development.

FIGURE 11.3. Remnant woodland vegetation has been retained within farmland pastures, but at a much lower tree density than prior to the European settlement.

January, and July winter temperatures are down around 5°C. Higher peaks in the mountains of the ACT can be snow covered in winter. Annual rainfall averages about 625 mm, with most rains occurring in spring and summer.

- Vegetation is dominated by eucalypt woodlands and open forests. (See remnant vegetation patches in figure 11.3.) In most cases, eucalypt trees are totally cleared or a few isolated trees are left in the landscape. (See foregrounds in figures 11.1 and 11.2.)
- Soils are formed from granites and metasediments derived from sedimentary rocks such as sandstones and siltstones. These natural soils

are of low natural fertility, but former farmland soils can have higher fertility due to previous applications of fertilizer.

Step 1: Setting goals

The first step in our adaptive landscape restoration process is setting clearly defined and measureable goals. (See chapter 1.) So that former farmlands would be more like the original grassy woodlands, ACT community groups set two goals: (1) to restore a more natural vegetation composition, they would plant iconic tree species (e.g., eucalypts); and (2) to provide appropriate habitat for native fauna, they would plant vegetation having the structures and pro-

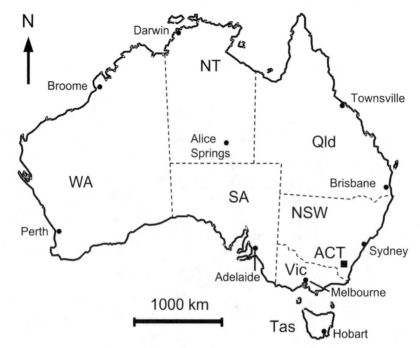

FIGURE 11.4. Location of the Australian Capital Territory (ACT) within New South Wales (NSW).

ducing the flowers and fruits required by particularly bird species. (See plate 11.1.) Although habitat restoration was specifically focused on vegetation and birds, the community groups also deemed the presence of reptiles, frogs, bats, and particular insects (moths) as important to the people utilizing the restored green space (figure 11.5).

When taking on all these challenging goals, ACT community groups also recognized that achieving landscape repair would take a considerable amount of time and that progress was likely to be incremental depending upon the availability of physical, financial, and human resources.

Community groups were concerned about other issues and aimed to control and eliminate soil erosion, such as the sheeting and gullying that had damaged the former farmlands (figure 11.6). They also intended to eliminate the aggressive weed species growing in gullies and in patches where remnant trees provided shady "camps" for sheep and kangaroos (figure 11.7), whose dung, in turn, increased soil fertility.

Step 2: Defining the problem

To achieve their goals, the second step in the restoration process was to critically analyze the situation, in this case eroded and weedy former farmland. The

FIGURE 11.5. Woodlands near ACT urban areas are used as green space by groups of bushwalkers and field naturalists.

Figure 11.6. An example of an eroding and weedy drainage line on a hillslope pasture in former ACT farmland.

Figure 11.7. Weeds colonizing brush packs built on the hillslope where trees have been killed by "camping" animals (livestock and kangaroos).

community groups faced a range of landscape factors that needed to be appreciated before embarking on restoration activities. We present these factors as being either positive (landscape properties that tend to facilitate restoration) or negative (properties or circumstances that make restoration difficult). We also note: *The more thorough the effort put into gaining knowledge about the problem, the better the chance of coming up with a successful restoration design.*

Positive factors

- Many desirable ground flora species were still present, though sparsely represented in the landscape and threatened by grazing rabbits and kangaroos.
- Although woodland trees were very sparse on the damaged pastureland and their seedlings and saplings were absent, community groups found that native tree species were locally available at plant nurseries.
- Although eroded in some places, the original topsoils were mainly still present. For groups designing restoration projects, this meant that sites would have some residual soil fertility and resistance to erosion.
- Some erosion features, such as gullies, were showing signs of self-repair (figure 11.8).
- Most former farmland areas in the ACT had some official protection status, so restoration activities would not be threatened by other land uses. However, this protection status also meant that community groups had to obtain permission from ACT government departments for their restoration activities, which could cause delays.

Negative factors

- The original clearing for farming was more than 150 years ago. This meant that community groups had to infer the original vegetation structure and species composition from nearby undisturbed landscapes.

- To establish plants that survive in former farmlands modified by fertilizers and, in some cases, cultivation, community groups had to learn the basic biology and ecology of the natural woodland plant species, which is not well understood.
- Because of stripping of topsoil layers, they found that soil fertility was too low in some places, but also too high in other places due to past fertilizer applications and enrichment by dung at animal camps under remaining trees, which promoted weeds. Camps also suffered from soil erosion due to trampling and exposure of hard-setting subsoils (figure 11.9).
- Because of drought periods, they knew that successfully establishing trees was risky.
- Because of steep slopes and fences on many hillslopes, they also knew that the use of machinery to assist with restoration activities was limited.
- Because of limited availability and expense, they found that some materials they would have liked to apply in restoration projects (e.g., commercial mulches) were too expensive.

Given all these factors, it was difficult for ACT community groups to decide which problems were most urgently in need of attention and where in the landscape to start their restoration activities. However, the five-step adaptive landscape restoration procedure, with its underlying framework and principles, provided them with a practical guide.

Steps 3 and 4: Designing solutions and applying technologies

A key restoration principle is to restore those landscape processes that are currently most dysfunctional and that have caused the most serious problems. In this case, excessive flows of runoff down hillslopes of former farmlands had cut gullies, and sheet erosion was continuing to strip away valuable topsoil around grass plants—a process called *pedestalling*. (See figure 11.9.) Essentially, if these landscapes were to be successfully converted into valuable green

FIGURE 11.8. An ephemeral drainage line showing signs of self-repair (grassy rounded walls and floor).

FIGURE 11.9. Soil-surface erosion on a hillslope caused by kangaroo grazing. Surviving grass plants have been pedestalled (top of photograph).

space, the remaining topsoils, as well as vital water from rains and runoff, had to be retained on hillslopes.

As part of their problem-analysis step, community groups identified and mapped the areas where excessive runoff had occurred on the hillslopes. They also identified probable run-on areas that were suggested by the presence of small patches of healthy remnant vegetation. Using this information, they designed and applied structures to first regulate flows of runoff water, that is, they applied two technologies to first enhance physical processes to retain water in landscapes. (See principle 2 in chapter 3.)

1. Logs barriers were positioned across hillslopes (figure 11.10) targeting those areas having excessive runoff rates and obvious sheet erosion. By first placing logs high on the hillslopes, they effectively intercepted and slowed overland flow at its source.

Zones just upslope of these logs became run-on patches where litter and sediments accumulated (figure 11.11).

2. Brush packs composed of woody branches were also constructed on hillslopes (figure 11.12). Commencing high in the landscape, brush packs were built just upslope of points where sheet erosion appeared to begin. This is where it is most critical to initially control overland flows. Brush packs were then constructed progressively farther down the slope. Brush packs are known to be very effective in arresting soil losses due to sheeting and, like log barriers, they eventually became patches of resource accumulation. (See Tongway and Ludwig 1996.)

Second, community groups applied technologies to enhance biological processes on hillslopes. (See principle 2 in chapter 3.) They established vegetation by planting native trees and shrubs (as tube

FIGURE 11.10. Logs and other coarse woody debris have been placed high on the hillslope to obstruct overland flows.

FIGURE 11.11. The upslope side of logs (to the right) accumulated litter and soil, but did not protect plants from being grazed.

FIGURE 11.12. Brush packs constructed on the slopes of former farmland pastures to obstruct sheet flows.

stock) in resource-rich run-on patches above log barriers (figure 11.13) and desirable trees and shrubs within brush packs. Although the branches provided some protection from browsing, plant guards were used to provide seedlings additional protection (figure 11.14).

To promote the success of these plantings, community groups also designed strategies to control weeds. First, weeds were evaluated relative to their aggressiveness and invasiveness, and also for their role in preventing erosion. Highly aggressive and invasive weeds were considered undesirable, but on these damaged pasturelands they also played a role in regulating runoff. This presented a dilemma to the community groups, which they solved with the following strategies:

- They removed patches of weeds known to be very aggressive and invasive, but to control erosion in these patches, they immediately applied other technologies such as brush packs.
- They did not remove weeds known to be less aggressive and invasive in areas where they were functioning to control overland flows. This allowed labor and physical resources to be freed up to focus on other areas with active erosion.

To control hillslope runoff and erosion, community groups chose not to use machinery to apply soil-surface ripping and banking techniques because these were too expensive and difficult to apply on steeper slopes. The application of such mechanical procedures may have also exposed dispersive subsoils, which would have inadvertently opened a Pandora's box of erosion problems.

Step 5: Monitoring and assessing trends
Because community restoration projects are gradually implemented over time, small adjustments to the procedures can be made in response to signals from monitoring previous restoration actions. These adjustments are an integral part of our adaptive landscape restoration procedure.

FIGURE 11.13. Trees and shrubs planted as tube stock in resource accumulation patches.

FIGURE 11.14. Tree seedlings with "plant-guards" planted within brush packs.

Early in the restoration process, community groups monitored and evaluated the logs and brush packs to determine if they were functioning

- To intercept overland flows and reduce sheet erosion on slopes below treated areas;
- To accumulate litter produced by trees, shrubs and grasses, an important indicator of soil quality improvement;
- To capture seeds that then germinated and established new plants, which, in turn, flowered and set seeds, which is a positive sign that the vegetation will become self-sustaining.

To serve as reference sites, community groups found inaccessible areas high in the landscape that had retained a dense cover of grasses and shrubs (figure 11.15). They assessed soil-surface condition indicators in these sites (see chapter 14) and also in areas where log-barrier and brush-pack treatments had been applied. They compared three condition indices: surface stability, infiltration capacity, and nutrient-cycling potential, and found that values on treated areas were below those expected from data collected at reference sites (figure 11.16).

Community groups also monitored hillslope drainage lines (see chapter 15) to see if sediments were still being stripped from the walls and floors of drainages once logs and brush packs were placed upslope to reduce over-bank flow rates, and if plants were colonizing and stabilizing ephemeral drainage-line walls and floors.

To determine whether vegetation on treated areas was being excessively grazed and browsed, community groups monitored native and feral herbivores (kangaroos, rabbits). Evidence of damage to new seedlings was also evaluated. Complete elimination of grazing and browsing was considered impossible, but regulating herbivores to reduce the proportion of foliage consumed was deemed important for maintaining plant vigor.

Community groups were particularly concerned with monitoring improvements in the structural complexity of vegetation (see chapter 16), which

FIGURE 11.15. Dense groundcover on an upper hillslope within a former farmland pasture, which serves as a reference area.

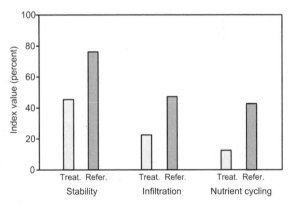

FIGURE 11.16. Values for three soil-surface condition indicators measured on treated areas (Treat.) compared to values for a reference area (Refer.).

indicates the capacity of the landscape to provide the habitats required by a greater diversity of fauna, particularly for birds. They knew it would take a number of years to fully achieve their goal of repairing hill-slopes so that they provide more habitats for a more diverse bird fauna. But, they were confident that applying restoration technologies, such as building brush packs and planting trees in these favorable patches, would eventually increase the structural complexity and functionality of hillslope vegetation; these habitat improvements would increase faunal diversity. In time, they would say that restoration trends were OK and that their goals had been satisfactorily achieved.

Further Thoughts

We want to emphasize that most of the people making up community groups are usually volunteers willing to spend their own time and money to restore former farmlands near their suburbs. This is fortunate because there is typically not much government funding or corporate assistance available to support landscape restoration projects. We have found these volunteers highly motivated people who have strong personal ethics to conserve flora and fauna; they develop a strong sense of a group responsibility for their restoration projects.

Community groups often take on projects to reverse declines in threatened species, such as song birds. To improve habitats for threatened species, community groups bring together individuals from all walks of life. To assess, for example, indicators such as bird population dynamics, experts are usually available within the community (e.g., ecologists). Because some community groups tend to have a narrow iconic species focus, a challenge for restoration ecologists working with these groups is to expand their appreciation of the ecology of landscapes so that their restoration goals and activities are appropriate to their capacity to *make a difference*.

Chapter 12

Restoring Verges after Road Construction

In many places around the world, roads bombard our senses with unattractive and unnatural landscapes, especially during and immediately after construction. To improved the aesthetic value of roads, RPs need to establish attractive vegetation along road verges. (See figure 12.1.) So as not to threaten the stability and safety of roads, RPs must also design ways to ensure that sediments from erosion of elevated road banks are not deposited on roadways. Runoff water is frequently concentrated in an erosive stream because it is harvested from a wide area and discharged over a single point on a road embankment.

In the construction of new, and repair of existing, urban and country roads, some roads may cut

FIGURE 12.1. A road verge where native grasses, mat rushes, and shrubs have successfully established and are providing an attractive and protective roadside covering. After rainfall events, no sediments accumulate at the foot of this embankment.

into hillsides, revealing the country rock, and others may be constructed on newly elevated landforms from rock and soil materials cut from the nearby landscape. But, in each case, water flowing over the new, and often steep, slopes has a high capacity to cause serious erosion and pollute adjoining streams and rivers. RPs face significant challenges preventing this erosion and pollution, while at the same time creating stable and attractive landscapes on road verges and embankments. (See figure 12.1.) In this chapter, our aim is to help RPs meet these challenges by putting into practice our restoration procedures and principles. Our scenario is based on our evaluations of road verge restoration and on the work of colleagues on highway rehabilitation. (See, for example, Ludwig and Courtenay 2008.)

Setting the Scene

Roads are obviously constructed around the world in all kinds of climates, vegetation, and soils. In this scenario we will assume that the road verge being restored was located in a region with the following attributes:

- A temperate, continental climate, where winters are cold and rainy (with some snow), and severe rainstorms are common during warm to hot, generally dry, summers.
- Mountainous terrain, where soils are rocky, mostly shallow, and moderately fertile.
- Forest vegetation, where tree canopies are up to 40 m in height, and the understory is typically grassy, but shrubs are common along rocky outcrops.

We will also assume that the road verge restoration work was the responsibility of an RP who is a landscape ecologist and part of a team within a state highway or roads department. Because the team included engineers and other personnel responsible for road design and construction, verge

restoration was a component of the planning process from the start.

Step 1: Setting goals
A new road, which would run through forest around the base of a low mountain and connect two suburbs, was proposed. During the planning stage for the construction of the road, the RP worked with the planning team to establish verge restoration goals. Their primary goal was to cover the newly constructed verge surfaces with vegetation and protective materials to achieve the following:

- Runoff from hillslopes in the surrounding landscapes and the verge would not threaten travelers on the road.
- Runoff sediments from road verges would not erode out or plug up constructed road drainage structures and contaminate streams below the road.

To keep future road verge maintenance to a minimum, they also set out to establish self-sustaining vegetation and to use stable protective coverings such as rock and woody mulch.

Because stretches of road ran through urban and suburban developments, they also needed to plant vegetation and use coverings pleasing to the eye.

Step 2: Defining the problem
The RP, in collaboration with other members of the team, conducted a number of assessments and analyses during both the planning and construction phases. The goals of restoration involved providing surface protection by covering some sections of road verge with coarse rock and mulch materials and other sections with soil and vegetation. They addressed the problems related to these two types of road section coverings.

Rock and mulch-covered sections

- To avoid the use of rock coverings that would readily breakdown or form acidic runoff when exposed to rains, they analyzed the physical and chemical weathering properties of all rock and stone materials proposed as protective coverings for road verges.

- To protect road verge surfaces from erosion using mulch, they evaluated the long-term resistance of proposed mulch materials to physical, chemical, and biological breakdown. They also evaluated the capacity of mulch materials, such as coarsely chopped wood-chips, to resist overland flows. In most cases, they found that this information was available from previous applications to road verges.

Soil and vegetation-covered sections

- To ascertain whether tree roots would be able to penetrate into rock fissures, or not, they investigated the nature of natural rock fracture patterns.
- To use soils suitable for establishing and supporting vegetation in the long term, they either used available topsoils or analyzed any soil materials proposed for use on road verge surfaces. They tested soil properties such as fertility (available nitrogen and phosphorus content) and stability (slaking and dispersivity). (See chapters 14 and 15.) Their aim was to avoid the use of low-fertility and unstable soils or, if their use was unavoidable, they wanted to know what amelioration treatments to apply (e.g., gypsum).
- To suit the rooting characteristics of intended vegetation plantings, they evaluated the availability of suitable soil materials to ensure that appropriately deep soil layers could be applied. They knew that grasses are usually shallow rooted, and most shrubs and trees are deep rooted.
- To cover open areas between clusters of plantings, they tested the longevity of coarse rock and woody mulch materials. They found that much of this longevity information was already available from prior road verge applications.

Other challenges faced by the RP and the team included the following:

- So that drainages could be designed to avoid concentrating flows onto road verges or onto the road surface itself, they needed to know how watersheds affected the roadway.
- To design roadway drainages and verge coverings that cope with runoff from roadways during rainfall events, they analyzed rainfall amounts and intensities for the region. They found that most rainstorms are brief, but can be highly energetic.
- To apply rock and mulch materials to road verge slopes too steep for machinery to negotiate, they explored ways of adding them from either the foot or the top of embankments.
- To avoid damage to roadways, they designed structures such as earth and rock banks to divert runoff. In some places, they also designed diversion banks to spread runoff water onto planted vegetation to help sustain it.

Steps 3 and 4: Designing solutions and applying technologies

The RP worked with the planning and construction team to design protective coverings of soil and vegetation, and rock and mulch, for different sections along the roadway.

Soil and vegetation-covered sections

The RP examined a variety of designs and applications for use on soil and vegetation covered road verges:

- If soil materials available for use as coverings had adverse properties such as being mildly dispersive, which could trigger erosion, the RP applied amelioration treatments such as gypsum.
- If road verges had steep slopes, the RP applied fabric mesh, enclosing mulch and grass seeds as a covering to protect soils and to grow vegetation.

- If soft materials such as mulch and soils were applied to areas next to those covered with harder materials such as rock, the RP designed flow pathways so that runoff from rock-covered areas did not damage areas covered with soil and mulch.
- To cover gently sloping verges with protective vegetation, after suitable soil materials had been added, the RP selected plant species adapted to the applied soils and to the climatic regime. The RP also considered other attributes, such as plant life form (tree, shrub, herb), mature structure (low, tall), longevity (short, long), and the need for ongoing maintenance. The RP selected a mix of plant species with these attributes, mostly locally sourced native plants. However, the RP did not exclude the use of exotic plants known to be well adapted to the local climate and soils.

The RP also selected different modes of planting vegetation, which depended on the terrain:

- On steeper slopes, the RP applied grass seed within mulch-fertilizer blankets.
- On more gentle slopes, the RP sowed grass seed directly into tilled soils.
- On slopes where shrubs and trees were designed to be in patches, surrounded by other coverings, the RP planted nursery-grown seedlings.
- For roadway locations with high visual exposure, the RP selected plant species with foliage and flowers that appealed to the eye and were attractive to birds and other fauna, such as butterflies.

Rock and mulch-covered sections
On steep verge slopes, the RP applied coarse rocks (about 5 to 25 cm in diameter), because they would be robust to raindrop impacts, overland flows, and weathering. On more gentle slopes less prone to erosion, the RP applied smaller rock and organic mulch materials known to be resistant to weathering.

Other design and application considerations
As noted earlier, avoiding excessive flows of water from hillslopes adjoining the roadway is a crucial design component, especially in preventing damage to roadway drainage structures and pollution of downslope creeks and streams during intense storm events. In this case too, the RP relied on hydrology experts on the team to design and engineer diversion structures on hillslopes above the roadway. Where needed, these experts designed and constructed water diversion banks above road cuttings to prevent severe erosion problems. (See figure 12.2.)

Another important roadway consideration is keeping larger wildlife species (e.g., livestock, horses, deer, and kangaroos) from wandering onto the road.

FIGURE 12.2. A road verge slope showing signs of sheet and rill erosion caused by excessive overland flows from above the roadway. Flows have also washed away the mulch that was originally applied to this slope.

The team designed a system of fences to exclude these large animals and also consulted with wildlife experts on ways to enable smaller wildlife species (e.g., possums, echidnas) to move between landscapes on either side of the roadway. They applied technologies such as the strategic placement of tree-top bridges and under-road chutes.

The RP also evaluated ways of controlling weeds that always appear along newly constructed road verges. They developed a system of timely spot-spraying newly emerged weeds with appropriate herbicides.

Step 5: Monitoring and assessing trends
The RP established a protocol for routinely monitoring road verges and steeper embankments and cuttings:

- To detect signs of erosion, the RP examined the upper and lower edges of road verges and looked for deposits of sediments at the bottom of cuttings and embankments and, in more severe cases, rills and gullies on verge slopes. If found, the RP looked for the cause of these erosion features. For example, sediment deposits and rills would likely be caused by focused overflows from hillslopes above the road during significant rainfall events. These overflows require corrective actions by hydrologists and engineers on the team.
- To evaluate the stability of verge surfaces covered with rock and mulch materials, the RP examined whether mulch remained in place after intense rainstorms. If mulch remained in place (figure 12.3), this indicates that it is providing a stable protective covering. However, if mulch on the verge surface had been washed or blown away (figure 12.4) then surface protection has been lost, and the surface is prone to erosion, particularly where raindrop impacts

Figure 12.3. Coarse mulch material has remained intact on a verge slope, protecting it from the erosive forces of rain and wind storm events. This mulch was sourced from when the forest was cleared to construct the roadway.

FIGURE 12.4. An area where mulch has been blown and washed off a verge surface leaving it unprotected. The soil now has an impermeable surface crust.

have crusted the surface. The result would be high runoff rates. In this case, the RP took remedial actions, such as applying a coarser mulch or rock material.

- To assess how well vegetation had established on planted sections of road verge, the RP looked to see if plantings had taken hold and were serving to protect surfaces and provide an attractive look. (See figure 12.1.) The RP also assessed if surface coverings (mulch, stone) in the spaces between plants were stable (figure 12.5). However, if the RP found that road verge planting had largely failed and that verge surfaces were showing signs of erosion, such as sediment and litter wash (figure 12.6), then the RP took corrective actions, such as using different plants or replacing plantings with rock and mulch coverings.
- The RP also looked along verges for any dead plants needing replacement or weeds requiring control treatments.

After a few years of monitoring and assessing the full extent of the rehabilitated road verge, the RP's overall evaluation was that trends were OK. (See figure 1.1.) However, the RP continued periodic monitoring to look for any problems such as vegetation dying, mulch washing off slopes, or rills forming. If any of these problems were found, or if members of the local communities reported such problems, the RP adaptively selected technologies to fix them in a timely manner.

Further Thoughts

In this chapter, we did not cover all the scenarios that RPs encounter when rehabilitating road verges. For example, to protect roadways from falling rocks where cuttings expose mountain rocks to weathering, engineering solutions such as steel netting or concrete coverings are needed. These technologies are beyond the scope of this chapter. Fortunately,

FIGURE 12.5. A protective wood-chip mulch and stone covering between plantings.

FIGURE 12.6. An "island" road verge where a surface mulch of fine stones has failed to resist erosion by overland flows. Note the wash of surface litter and sediments.

most RPs are members of a team that has engineers expert in designing protective coverings for rock faces through road cuttings, so that these RPs can focus on designing coverings to protect more gently sloping road verges from erosion. Our scenario is aimed at RPs who use coarse rock, stones, coarse organic mulch, soil, and vegetation to rehabilitate road verges, and who will be successful by putting landscape restoration procedures and principles into practice.

PART IV

Monitoring Indicators

In the chapters of part 4, our aim is to describe for restoration practitioners those indicators we have found useful for monitoring and assessing landscape function, that is, how well landscapes are working as systems to regulate resources and provide goods and services. Earlier in our book (chapters 1 to 3 in part 1), we defined indicators as easily observed surrogates of difficult-to-measure attributes. Here we add our view that, while there are numerous indicators that can be used to evaluate landscapes, we have found that many of these indicators only reflect changes in ecosystem structure and composition, such as the loss of plant species, and do not directly indicate processes. Our years of experience in evaluating landscape restoration have led us to the conclusion that measures of vegetation composition by themselves do not provide the information about critical processes that restoration practitioners need to design effective restoration technologies. For example, what does the absence of a species really mean? Rather, we have found that designing successful restoration technologies requires identifying, assessing, and analyzing the functional processes that are declining in the landscape, and the causes of these declines.

To monitor and assess indicators of functional processes, we describe in the four chapters of part 4 a set of methodologies known as *landscape function analysis,* or LFA for short. In chapter 13, we first provide restoration practitioners with an overview of LFA, and then we describe those LFA indicators that we know they will find useful for characterizing how landscapes are structurally and functionally organized. In chapter 14 we describe LFA soil-surface indicators that define functional processes operating in a landscape, such as the capacity to resist erosion. In chapter 15, we described indicators that restoration practitioners can use to define the condition of ephemeral drainage lines running through a landscape. In chapter 16, we define those measures of vegetation structure and composition that indicate the capacity of landscapes to provide a complex of habitats for a diverse biota.

These four chapters provide restoration practitioners with a full range of LFA indicators and monitoring methods that enable them to design and apply effective restoration technologies. Most important, these methods and indicators provide them with the data they need to track progress toward desired landscape restoration goals and, if necessary, adapt technologies to improve progress.

Chapter 13

Landscape Function Analysis: An Overview and Landscape Organization Indicators

In this chapter our aim is to first provide restoration practitioners (RPs) with an overview of landscape function analysis (LFA) and then with a description of those indicators that they can use to characterize how a landscape is structurally (spatially) and functionally organized. Here we only briefly describe these landscape organization indicators (LOI) because detailed descriptions are provided in a LFA LOI document available online (http://members.iinet.net.au/~1fa_procedures). Also available online are LFA field data sheets and spreadsheets we refer to in this, and later, chapters.

Overview

LFA was originally developed to establish a set of soil-surface indicators for measuring and analyzing the nature and severity of problems in dysfunctional rangelands. (See Tongway 1995; Tongway and Hindley 2000.) Over the years the scope of LFA has greatly expanded to include methods useful for monitoring restoration trends in many different types of landscapes. (See Tongway and Hindley 2004.)

LFA is composed of three modules: (1) a conceptual framework, (2) indicators of landscape function and field procedures for monitoring these indicators, and (3) an interpretational framework.

Module 1. Conceptual framework
We described the conceptual framework underlying LFA in chapter 2. We used a series of diagrams and

photographs (figures 2.1 to 2.12) to illustrate this framework and to explain how landscapes function as biophysical systems. The framework emphasizes the sequence of input, internal transfer, and output processes operating to regulate flows of energy and materials such as water, soil sediments, and organics (e.g., litter, seeds). This framework helps remind RPs of the need to maintain both spatial organization and processes. In chapter 3 we described four principles central to this conceptual framework that, when put into practice, will help RPs repair damaged landscapes.

Module 2. Indicators of landscape function
In this module we describe indicators reflecting the state of functionality of landscape processes and ways for RPs to assess these indicators in the field. As noted earlier, methods for assessing specific indicators are available as online documents, and spreadsheets are also included online to help RPs reduce raw data to sets of indices. All of these online LFA documents are updated as methods are improved and provide RPs with details on monitoring procedures and measurements, which are illustrated by numerous diagrams, tables, and color photographs.

In part 4, we use a selection of these diagrams, tables, and photographs (in grayscale) to describe landscape function indicators in four chapters:

- To indicate how landscapes are spatially organized, we describe a number of useful site attributes here in chapter 13.

- To indicate landscape surface stability and the potential for landscape systems to cycle nutrients and infiltrate water, we describe eleven soil-surface condition indicators in chapter 14. These eleven indicators are then combined into three synthetic indices: surface stability, infiltration capacity, and nutrient-cycling potential.
- To evaluate the stability of ephemeral drainage lines, such as gullies running through landscapes, we describe eight drainage-line indicators in chapter 15.
- To evaluate different functional attributes of vegetation, in chapter 16 we describe methods for estimating vegetation species composition, horizontal and vertical structure, and habitat complexity.

Module 3. Interpretive framework

Typically, restoration indicator data are collected at two scales: a coarser watershed-hillslope site scale, and a finer vegetation-soil patch scale. Our interpretive framework primarily aims to inform RPs about the functional status of a landscape at the site scale. It does this by comparing values for indicators measured on rehabilitated sites with those measured on reference sites. These comparisons are essential because they enable RPs to evaluate and judge how well rehabilitated sites are progressing toward established goals.

Here we briefly outline a few key features of this LFA interpretive framework, but a more complete description is provided online (http://members .iinet.net.au/~lfa_procedures). Progress of indicators toward landscape restoration goals can be illustrated and interpreted with two basic types of graphs that (1) plot trends in indicators along one-dimensional lines or continuums, and (2) plot responses of indicators over time in two dimensions.

The first graphical procedure was illustrated by landscape restoration principle 4 and figure 3.6 back in chapter 3; these illustrations were about the importance of assessing the progress of indicators of landscape function along a line (continuum) toward greater functionality, a restoration goal. In this graphing procedure, measurements for individual indicators obtained at different monitoring times are plotted as time-marks along a continuum line, which extends from indicator values representing dysfunctional landscapes to values for highly functional landscapes. These time-marks are informative because they represent significant and specifiable changes in landscape functionality along a continuum.

The second graphical procedure simply plots data for indicators being monitored as changes over time (time-trace graphs). For indicators measured on rehabilitating sites, these time-trace graphs typically have an S-shaped or sigmoidal form because many landscape processes develop slowly at first and then accelerate before eventually leveling off. Recall that these S-shaped curves were typical of many of the indicator data we plotted as time traces in our case studies and scenarios. (See chapters 4 to 12.) In some cases the indicator increases rapidly from the start and then gradually levels off. Such curves can be described as logarithmic or as an exponential rise to a maximum.

Properties of S-shaped, logarithmic, and exponential response curves can also be used to estimate thresholds, for example, the point at which response data for an indicator of a biophysical process suggests that the landscape is becoming self-sustaining. Thresholds are useful both conceptually and in practice. (See Du Toit et al. 2003.) Threshold points indicate when the resilience of landscape functioning is sufficient to cope with normal disturbances without significant loss in landscape functionality. For an indicator with an S-shaped response curve, this threshold point can be taken as the inflection point, which is the point along an S-shaped curve where the rate changes from increasing to decreasing, that is, the rate begins to slow as it trends toward leveling off. This upper level represents the most functional state a landscape can be in, given the soil type, landform, and climate. In a sense, this upper level is the landscape's biogeochemical potential.

An alternative way of estimating a threshold along an S-shaped or logarithmic response curve is the halfway point, which is based on the highest and lowest values measured for an indicator:

Threshold = [(highest value–lowest value)/2] + lowest value

Thresholds estimated by halfway points have been useful for interpreting trends in LFA indicators on sites being restored from gold mine tailings. (See Haagner 2009.)

Finally, if you look back at figure 3.1 in chapter 3, you will see that S-shaped curves (increasing and decreasing forms) are used to interpret the progression of biological processes taking over from physical processes. This takeover is a core principle signifying restoration success. After applying restoration technologies, RPs can use this principle to better interpret to what extent they are achieving their restoration goals by tracking the progression of indicators reflecting both physical and biological landscape processes (e.g., water retention, vegetation growth).

Landscape Organization Indicators (LOI)

We start our description of how to describe indicators of landscape organization by listing what tools and equipment are needed to conduct field measurements.

Field Equipment

- 100 m tape measure and steel pegs to secure the tape in position
- 10 m hand-held tape measure
- Stainless steel paint scraper with a 60 mm wide blade
- Data sheets and clipboards
- Two liters of rainwater or equivalent
- Small shallow dish with a diameter of about 100 mm
- Copy of the LFA *Field Procedures* manual

Measurement Scales

As noted earlier, the RP uses this equipment in the field to collect data on indicators of landscape processes at two scales: the site scale (e.g., hillslope lengths of 100 to 200 m or more), and the patch scale (e.g., vegetation patches of 1 to 2 m in diameter located within sites).

Coarser site scale methods describe how to measure indicators of landscape organization, which enables RPs to assess the fate of mobile resources (e.g., water, sediments, litter, seeds). In other words, RPs use these indicators to discover whether mobile resources are internally redistributed, retained, and used within the landscape, or are transferred off the site to other landscape systems. The fate of mobile resources is indicated, for example, by measuring the spatial attributes of vegetation on a hillslope. These spatial attributes indicate the capacity of the landscape to obstruct and absorb overland flows of water and ameliorate wind erosion. Such indicators include the number and size of perennial plant patches (e.g., grass clumps, shrub thickets) where resources on the site tend to accumulate, and the vertical distribution of plant canopies (e.g., foliage density) where fauna such as birds and lizards tend to concentrate.

Landscape organization data, and the soil-surface condition data described in the next chapter, are collected on line transects. These transects are oriented in the direction of water flow (downslope) or, if wind is the force of primary concern, downwind. These line transects are called gradsects, short for *gradient-oriented transects* (Gillison and Brewer 1985), and the data have direct cause-and-effect spatial relationships. Gradsects are a very efficient means of collecting landscape function data. Many other site assessments use random or systematic grid-based sampling procedures; however, winds and flows of water are not random processes in landscapes, and defining the nature and scale of the emergent heterogeneity (pattern) is crucial to understanding how landscapes function. (See chapter 2.)

At finer patch scales, soil-surface assessments (SSA)

are conducted to infer the activity of landscape processes such as soil erosion, water infiltration, and nutrient-cycling, and in doing so, add a quality dimension to the coarser site scale data. At the finer patch scale, eleven readily observable SSA indicators are assessed in representative patches and interpatches. (See chapter 14.) These patches and interpatches are defined along the landscape organization gradsects. For example, resources tend to be lost from bare interpatches and accumulate within perennial vegetation patches. We selected the eleven SSA indicators to provide RPs with information on a wide range of readily observable landscape processes that can be validated by field and laboratory measurements.

As noted earlier, a spreadsheet program is available online to integrate the eleven SSA indicators across site and patch/interpatch scales to provide information that RPs will find useful for interpreting findings at the broader site scale. The LFA monitoring procedure is, therefore, able to coordinate observations made at different scales to provide RPs and stakeholders with insights into the degree of function/dysfunction of landscape processes.

The Site

To establish the context for a site being rehabilitated, we recommend that RPs describe its underlying geology, slope, aspect, vegetation, soil type, and current land use. (See table 13.1.) To determine where in the landscape processes are most active, and perhaps dysfunctional, RPs need to carefully and broadly examine the entire site. With this information, RPs can position gradsects to most efficiently measure (sample) indicators of landscape processes. These contextual attributes of the landscape are also used to select and gather information from reference sites located within the region.

Spatial Organization

A number of activities, in the order listed here, are used to characterize the spatial organization of landscapes on restoration and reference sites:

TABLE 13.1

LFA SITE DESCRIPTION
Site no: _____   Date: _____
Site name:
Observer(s):
Position (GPS):
Transect compass bearing:
Position in landscape:
Geology:
Soils:
Slope: _____   Aspect: _____
Vegetation type:
Land use:
State of soil surface:
Comments:

Example of an LFA data sheet used to characterize sites.

Activity 1: Position gradsects

Using the 100 m tape, gradsects are oriented directly downslope (bends in the tape may be necessary in some cases to follow the direction in which resources flow), or in the direction of the prevailing wind if this is the primary factor of interest. To facilitate repeated measurements over time the location of the gradsect is recorded and the site permanently marked. This is essential for long-term monitoring if trend assessments are to be meaningful. Typically, the gradsect location is documented from global positioning systems (GPS) readings at the upslope and downslope ends, and a compass bearing is also taken.

Activity 2: Define patches and interpatches

Starting at the top of the gradsect, a continuous record of patches and interpatches is recorded as line intercept distances between their respective boundaries (zones of resource accumulation versus loss). (See figure 13.1.) These patch/interpatch tabulations are completed before commencing tasks such as soil-surface assessments. (See chapter 14.) We find it useful to tabulate patch/interpatch data using the spreadsheet (available online) that automatically calculates patch and interpatch indicators of landscape functionality. See, for example, the patch and interpatch data in table 13.2. Sometimes a site

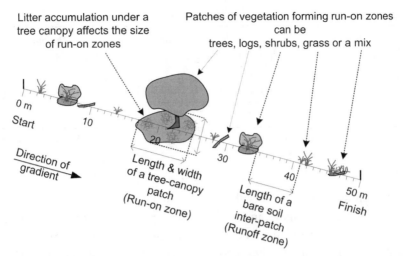

FIGURE 13.1. A diagram representing a 50 m gradsect with continuous recording of resource accumulating patches (run-on zones) such as grass clumps, shrubs, and trees, and resource shedding interpatches (runoff zones) such as bare soil.

TABLE 13.2

Example of typical landscape organization data collected along a 50 m gradsect. Distances defining the boundaries from the beginning to the end of patches and interpatches are recorded from the start and to the finish of the gradsect line. The type of patch or interpatch is identified in the row recording where it ends. These data map the way the landscape is spatially organized with respect to resource regulation.

Distance along tape measure (m)	Patch width (cm)	Patch/interpatch identification	Notes on types of patches and interpatches, and factors such as grazing
0			[start of gradsect]
2.5		BS	Bare soil (interpatch)
3.3	710	OT	Open shrub thicket (patch)
5.9		BS	
6.8	80	PH	Plant hummock (patch)
9.2		BS	
9.22	10	PH	(grazed to about half-height)
10.4		BS	
11.5	130	OT	(a few shrub-tips browsed)
13.32		BS	
15.34	10	PH	(grazed to near base)
15.9		BS	
16.4	105	SLC	Shrub log complex (patch)
21.15		BS	
21.4	30	PH	(grazed to quarter-height)
22.85		BS	
23.6	105	SLC	
33.35		BS	
33.6	35	SLC	
35.2		BS	
37.0	650	OT	
40.1		BS	
42.0	200	OT	(a few shrub-tips browsed)
50.0		BS	[finish of gradsect]

TABLE 13.3

Summary of data in table 13.2 on indicators of landscape organization. These site-scale totals and averages would be compared to values obtained over time to examine changes in landscape functionality.

No. patch zones per 10 m	Total patch zone width (m/10 m)	Mean interpatch length and range (m)	Landscape organization index*
2.4	4.1	3.2 (0. 6–9.8)	0.22

* Sum of all individual patch lengths measured along the gradsect (see figure 13.1) divided by the total length of the gradsect

may be classified entirely as a single patch, with no discernable interpatches as, for example, in dense perennial grasslands. Sometimes a site may be composed of alternating patches of different types, for example, dense grassy patches interspersed by woodland groves. The width of a patch is judged by observing where resources flow around the lateral boundaries of the patch.

We have found it useful to summarize patch and interpatch data. (See table 13.3.) In this example, 2.4 grass clumps per 10 m of gradsect provide 4.1 m of obstruction width to downslope flows. This indicates a high capacity of the landscape to capture resources being driven by water- and wind-mediated processes. This capacity is illustrated in figure 13.2. RPs can use such data to address a central question: Are mobile resources tending to be accumulated in patches or tending to flow around patches along open interpatches?

FIGURE 13.2. Clumps of grass plants facilitate the capture and accumulation of resources driven by both water and wind processes (depicted as arrows).

Activity 3: Measure reference sites
To assess whether rehabilitated sites are achieving restoration goals, RPs need to select reference sites and conduct activities 1 and 2 on these sites. Recall that reference site data provides numerical target values to assess restoration trends. (See chapter 3.) With data from reference sites, RPs can address the following kinds of questions:

- What patch size and density characterize the reference site? How does the restoration site data compare with these reference site data?
- Are interpatch lengths short enough to prevent excessive flow rates and to minimize erosion?
- Where in the landscape are lengthy interpatches located?
- Are patches becoming larger or more numerous over time? If so, at what rate?

Note that patch and interpatch quality are not directly addressed by indicators of landscape organization. This is the role of the indicators described in the next chapter.

Note: The *LFA Field Procedures* manual and its contents are the property of the Commonwealth Scientific and Industrial Research Organisation (CSIRO), Australia. The maintenance, updating, and online distribution of this manual, and other LFA documents referred to in this book (available at http://members.iinet.net.au/~lfa_procedures), is the responsibility of one of us (DT).

Chapter 14

Landscape Function Analysis: Soil-surface Indicators

Here we describe eleven indicators of soil-surface processes that restoration practitioners (RPs) assess in each of the patch and interpatch types identified along landscape gradsects. (See chapter 13.) These soil-surface assessments (SSA) are made on examples of each patch and interpatch type selected at random from the full set recorded along gradsects. We call these selected examples query zones. Typically, three to five query zones are assessed per patch or interpatch type to obtain statistical rigor. As a rule of thumb, if we find a considerable amount of variation between patch (or interpatch) query zones within a type, then we assess a greater number of zones. A more rigorous method for estimating the number of zones to assess is explained in an LFA document on sample size available online (http://members.iinet.net.au/~lfa_procedures).

Each query zone is identified by its location (distance) along the gradsect and the eleven soil-surface features, or indicators, are scored according to SSA methods described in an LFA document also available online (http://members.iinet.net.au/~lfa_procedures). RPs will find it useful to enter their assessed scores for the eleven indicators onto a field data sheet (table 14.1). These field assessment data should be entered into an SSA data entry spreadsheet, which is also available online (http://members.iinet.net.au/~lfa_procedures), as soon as possible (ideally, while in the field, but in any event, on the day of collection). We find that quickly entering data into a spreadsheet helps

TABLE 14.1

LFA Soil-Surface Assessment (SSA) Field Data Sheet

Date: _____   Observer(s): _____

Site Name: _____   Transect: _____

SSA Indicator (possible scores)	1.5 m BS	2.9 m OT	6.4 m PH	9.8 m BS	14.0 m PH	34.5 m PH	38.5 m BS	41.0 m OT
Rainsplash protection (1–5)								
Perennial vegetation cover (1–4)								
Litter cover, origin, & incorporation (1–10)								
Cryptogam cover (1–4)								
Crust brokenness (1–4)								
Erosion type & severity (1–4)								
Deposited materials (1–4)								
Soil-surface roughness (1–5)								
Surface resistance to disturbance (5–1)								
Slake test (1–4)								
Surface texture (1–4)								

At the top of each column, note location of the query zone along the transect (in m from 0) and the code for the type of patch or interpatch assessed:

Example of a field form for scoring eleven soil-surface indicators assessed at eight patch, or interpatch, query zones selected along a gradsect (e.g., BS—bare soil [interpatch], OT—open shrub thicket [patch], and PH—plant hummock [patch]) as listed at the top of the columns. Note that the specific location is recorded in case a repeat field visit becomes necessary.

verify whether field assessments of indicator scores are within valid ranges. If assessment errors are found, they can then be readily corrected while in the field.

Soil-Surface Indicators

Below we briefly describe the landscape processes assessed by the eleven soil-surface indicators. We also note whether each indicator contributes to one or more of the three synthetic soil-surface condition indices defined at the end of this chapter: the stability index, the infiltration index, and the nutrient-cycling index.

1. Rain splash protection

This indicator assesses the degree to which surface covers such as stones and perennial grasses, and low shrub and tree foliage, ameliorate the effects of raindrops impacting on the soil surface. Raindrops can cause soil-splash erosion and soil surfaces to form physical crusts. The rain splash protection indicator contributes to the stability index.

2. Perennial vegetation cover

Through aboveground measurements, this indicator assesses the contribution of the belowground biomass of perennial vegetation to nutrient-cycling and infiltration processes. As explained in the SSA document available online, the basal crown diameters on perennial grasses and the canopy diameters of shrub and tree canopies are measured within query zones. These query zone cover measurements are used to score this indicator, which contributes to both the infiltration index and the nutrient-cycling index.

3. Litter

Here, litter refers to the following materials in the query zone: annual grasses and other ephemeral herbage (both standing and detached); any detached leaves, stems, twigs, and fruit of perennial grasses, trees, shrubs, and forbs present; and any animal dung. These litter materials function to protect soil surfaces from rain splash, and hence, litter cover contributes to the stability index. The degree to which litter is being decomposed strongly relates to the concentration of carbon, nitrogen, and other elements stored in surface soil layers and soil-surface porosity, hence, the litter indicator also contributes to the infiltration and nutrient-cycling indices.

4. Cryptogam cover

For our purposes, *cryptogam* is used as a generic term to include cyano-bacteria, algae, fungi, lichens, mosses, liverworts, and fruiting bodies of mycorrhizae. When present, the cover of cryptogams significantly contributes to the stability of soil surfaces. (See Eldridge and Greene 1994.) Typically (though not exclusively) they colonize soils in open areas that have preexisting undisturbed physical crusts. Cryptogams usually need high light levels to persist and are seldom found under dense litter, particularly woody litter. They are, however, observed under thin layers of grassy litter. Cryptogams may be early colonizers of recovering soil surfaces but may decline as vascular plant cover increases.

Cryptogams impart flexibility and stability to surface crusts because their hyphae penetrate into the soil surface, and hence, their cover contributes to the stability index. They are also associated with elevated levels of available nutrients in the surface layers of soil, thus their cover also contributes to the nutrient-cycling index.

5. Crust brokenness

Here, we define soil-surface crusts as densely packed physical surface layers that overlie subcrust materials. Physical crusts in good condition are smooth and conform to the gentle undulations in the soil surface. Such crusts yield little soil material in runoff events but may restrict infiltration. However, crusts can become unstable, brittle, and easily disturbed by grazing animals. In this case, soil-surface materials becoming available for wind or water erosion, hence, this indicator contributes to the stability index.

6. Soil erosion type and severity

Soil erosion refers to those signs of active soil loss caused by water and wind action. (See figure 14.1.) In this context there are five distinct types of soil erosion. (See table 14.2.) These five types of surface erosion are based on descriptions in a handbook by McDonald et al. (1990). These types of erosion are also illustrated by color photographs in the online LFA SSA document. Assessments of soil erosion type and severity contribute to the stability index.

FIGURE 14.1. Five types of soil-surface erosion: sheeting, pedestal formation, rilling, terracette formation, and scalding.

7. Deposited materials

Recently deposited soil and litter materials found in a query zone indicate surface instability upslope or upwind of the zone. Surface instability permits loose materials to be transported into the query zone. These materials are usually silts, sands, and gravels (alluvium). Absence of these materials does not necessarily imply a lack of alluvial transport because water and wind erosion may actually sweep them completely off the query zone being assessed.

Assessing the amount or volume of deposited alluvium is more important than simply observing the presence of alluvial deposits, because these deposits may quickly become obscure due to germination of seeds within the deposit. For example, alluvial fans may stabilize and become covered with pulses of plant growth soon after rainfall events.

This material deposition indicator contributes to the stability index.

8. Soil-surface roughness

Surface roughness indicates the capacity of the soil surface to slow overland flows and to retain mobile resources such as water, seeds, topsoil, and litter. This indicator contributes to the nutrient-cycling index because it assesses the capacity of a query zone to capture materials. The roughness of a soil surface may be due to features such as depressions and cracks in the soil surface, which very effectively function to capture and retain mobile soil and litter particles from water runoff and wind. (See chapter 5.)

Another surface roughness feature is a high density of plants, such as perennial grasses, and plant litter beds. Basal butts of grass plants cause overland

TABLE 14.2

Five types of soil erosion

Erosion type	Description
(E) Sheet erosion	The progressive removal of very thin layers of soil across extensive, gently sloping areas, with few if any sharp discontinuities to demarcate them. Sheet erosion is not always easy to detect with assurance and may need to be inferred from other soil-surface features, such as coverings of gravel or stone (called "lag") left behind after erosion of finer material, or the presence of downslope eroded materials. Sheet erosion is sometimes confused with scald erosion, which typically occurs on flat areas with texture-contrast soils.
(P) Pedestals	The result of soil being eroded from around a plant to a depth of at least several cm, leaving the butt of the surviving plant on a column of soil above the new general level of the landscape. Exposed roots are a hallmark of this erosion form. This is a sign that the soil type itself is very erodible and that loss of vegetation in the landscape was preceded by erosion, not the other way around. Often associated with stones as the protective cover in the post-mining environment.
(R) Rills and gullies	Distinctive channels cut by flowing water. Rills are less than 300 mm deep, and gullies are greater than 300 mm deep. They may be initiated by water flowing down sheep or cattle paths. Their presence is a sure sign that water flows rapidly off the landscape, often carrying both litter and soil with it. They are aligned approximately with the maximum local slope.
(T) Terracettes	Abrupt walls from 1 to 10 cm high, aligned with the local contour. Terracettes progressively cut back upslope, the eroded material being deposited in an alluvial fan downslope of the feature. The location of a terracette should be noted in the comments of the landscape organization sheet for the line transect so that the progress of the terracette upslope can be monitored over time. A change of zone will occur at the location of the terracette, and it is assessed as occurring in the upslope zone. It will have an erosion type and severity class value of 1 or 2. The erosion type downslope of the terracette may be sheeting with alluvial deposits.
(S) Scalds	The result of massive loss of A horizon material in texture-contrast soils. Erosion exposes B horizons, which are typically very hard when dry and have extremely low infiltration rates. Scalds are typically found on flat landscapes; they pond or shed water readily and are bare of vegetation.

flows of water to be slower, deeper, and highly convoluted, sieving out any particulates. (See figure 2.1c.) The slowing of flow rates allows for longer infiltration times, hence, this indicator also contributes to the infiltration index. A high density of plants also provides safe sites for the lodgment of seeds and litter blowing in winds.

9. Surface resistance to disturbance (coherence)
Soil surfaces differ greatly in the strength with which they resist physical disturbance such as from the hooves of grazing cattle and feral animals or from vehicle or pedestrian traffic. Noncohesive soil surfaces are prone to such disturbances because surface particles are readily eroded by water and wind. In contrast, very hard soil surfaces have resistance to physical disturbance because of their high mechanical strength, but they also have a very low capacity for infiltration. This is due to their low porosity and

tendency for massive crusting or hard setting. This tradeoff is automatically taken into account by the online SSA spreadsheet program, which appropriately weights the stability and the infiltration indices by surface crust and surface coherence scores.

10. Slake test
Stable soil fragments maintain their cohesion when wet, indicating soils with a low potential for being eroded by water. As described in our case study and scenario chapters, determining the stability of materials to wetting (e.g., during rains, flooding, or irrigation) before using these materials in landscape restoration projects is very important. The slake test is used to assess scores for the stability of soil fragments to rapid wetting. This test is described in the online LFA SSA document. Scores from conducting the slake test contribute to the stability index and the infiltration index.

11. Soil-surface texture

Soil texture is an important indicator because it strongly affects infiltration and runoff processes, hence, this indicator contributes to the infiltration index. Assessing soil texture in the field is a bit messy. (See the online LFA SSA document.) Fortunately, this procedure is usually only conducted at the time of site establishment. It does not need to be repeated in each query zone, unless there are obvious changes in surface soil textures along the gradsect.

Synthetic Soil-Surface Indices

As noted in our descriptions above, the eleven soil-surface indicators are assessed and used in various combinations (figure 14.2) to derive three soil-surface condition indices: stability index (surface resistance to erosion); infiltration index (potential of surface soils to infiltrate water); and nutrient-cycling index (potential for surface soils to cycle nutrients back into the soil). These three indices are automatically calculated using equations programmed into the LFA SSA data entry spreadsheet. Each index has a nominal scale of 0 to 100 percent or, if preferred, 0 to 1. The higher the index number the higher the functionality of the landscape process (stability, infiltration, nutrient cycling). These synthetic indices relate to the landscape processes illustrated and described in the conceptual framework (see chapter 2). These indices have been verified against laboratory and field measurements covering a range of landscape types, climatic regimes, and land uses. A report on verification of LFA soil-surface indicators is available online (http://members.iinet.net.au/~lfa_procedures).

To produce scores reflecting the current overall state of functionality of the landscape, the spread-

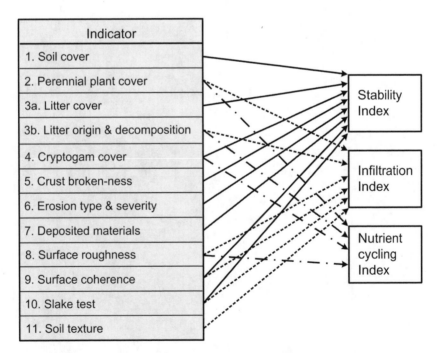

FIGURE 14.2. Different combinations of 11 soil-surface condition indicators are used to calculate three synthetic indices of the potential for the site to resist erosion (stability), retain and store water (infiltration), and cycle nutrients to enhance plant growth (nutrient cycling).

sheet program does calculations to integrate indicator data from the finer vegetation-soil patch scale and coarser watershed-hillslope site scale. To decide at which scale (patch or site) the most useful infor- mation emerges, RPs can evaluate soil-surface con- dition information and patch and interpatch data in relation to the framework on how landscapes func- tion. (See chapter 2.)

Chapter 15

Ephemeral Drainage-Line Assessments: Indicators of Stability

In chapter 15 we describe eight indicators that inform restoration practitioners (RPs) about whether a drainage line is still active or is becoming stabilized. Here, we are referring to drainage lines that flow occasionally, that is, ephemeral drainages, such as gullies running down a hillslope rather than permanently flowing creeks, streams, or rivers. Ephemeral drainage lines on hillslopes in undisturbed landscapes are smoothly concave in cross-section (no sharp edges), receive diffuse overland flow, and gently drain water from the surrounding slopes. (See figure 15.1.) These undisturbed landscapes do not have high rates of runoff or erosion. Landscapes that have a history of disturbance, however, often have hillslopes where drainage lines have become incised. (See figure 15.2.) If erosion cuts a channel deeper than 0.3 m, the drainage line is defined as a gully; if the cut is less than 0.3 m, it is called a rill.

Depending on soil type and the nature of disturbance, even initially shallow ephemeral drainage lines (rills) on gentle hillslopes can erode into deep

FIGURE 15.1. A hillslope on which the natural ephemeral drainage line shows no sign of an incised channel. Lateral water flows into the drainage line are diffuse and slow and the cross-section of the line is smoothly concave.

FIGURE 15.2. A channelized ephemeral drainage line on a hillslope shows signs of rapid over-bank flow (bare side-walls) but a stabilizing grassy floor. This implies that the restoration solution involves reducing flow rates from the hillslope above the channel, rather than within the channel itself.

gullies. (See figure 15.3.) Restoring even small but heavily eroded drainage lines presents RPs with major landscape restoration challenges, let alone those that are large, deep, steep-walled gullies. (See figure 15.4.) However, we know that RPs can successfully restore ephemeral drainage lines by putting into practice our five-step adaptive procedure and its underlying principles, because RPs met similar challenges in our case studies on restoring damaged rangelands. (See chapter 5.)

In chapter 15 we describe eight indicators of ephemeral drainage-line functioning. Our aim here is to briefly convey to RPs what landscape processes these indicators represent. Details of the methods RPs use to score these eight indicators in the field are provided in a document on ephemeral drainage-line assessment (EDA), which is available online (http://members.iinet.net.au/~lfa_procedures/). To help RPs score indicators, this document refers to color photographs illustrating ephemeral drainage lines in different states of stability.

If the eight indicators listed here are assessed using EDA methods, the RP can evaluate the urgency of restoring the ephemeral drainage line and can use landscape restoration principles and pro-

FIGURE 15.3. A gully 1.5 m deep has been cut into a hillslope ephemeral drainage line. Note the original position of the soil level at the root crown on a small tree.

Figure 15.4. A deep, steep-sided gully

cedures to design and locate appropriate treatments. Two of the eight indicators address characteristics of the slopes flanking the drainage line. The remaining six indicators focus on the drainage line itself. These six indicators assess the vegetation within the drainage line and determine the shape and erodibility of the channel floor and walls.

Assessing Indicators of Potential Runoff from Adjoining Slopes

The nature of the slopes above and along the ephemeral drainage line can strongly affect the rate and amount of runoff likely to impact the drainage line by over-bank flow during rainstorm runoff events. To assess these potential impacts of runoff, two indicators are scored: steepness of the slopes, and the amount of surface protection on the slopes.

1. Slope steepness indicator

The steepness of the slopes above and bordering the ephemeral drainage line affects the potential energy of any runoff flowing into the line during rainstorms. Basically, this indicator assesses the contribution of slope steepness to the potential for high flow rates into the drainage line and resultant erosion of the drainage channel walls and floor.

2. Slope-surfaces indicator

The amount and rate of runoff coming from above an ephemeral drainage line strongly affects its stability. This runoff is regulated by the amount of vegetation, litter, and coarse debris on the hillslope. Essentially, this indicator assesses the role that such surface protection materials play in the contribution of overland flows into a drainage line and potentially eroding it.

The above two indicators help RPs determine whether any ephemeral drainage-line erosion is being caused by flows of water from farther upstream in the channel or by water cascading over the lip of the drainage line from areas directly adjacent to it. (See plate 2.2.) Identifying the source of excessive overland flows can assist RPs in designing appropriate ephemeral drainage-line restoration technologies. This restoration design can also be aided by assessing soil-surface condition indicators along gradsects oriented upslope and adjacent to the drainage line. (See chapter 14.)

Assessing Indicators of Ephemeral Drainage-line Vegetation

Vegetation growing within an ephemeral drainage line that has, for example, eroded into a gully provides RPs with an indication of the potential for the gully to erode. Dense, long-lived, perennial vegetation within a drainage line indicates that the line has been stable for a substantial period of time and has resisted erosion during recent runoff events. (See figure 15.1.) Short-lived vegetation within the drainage line provides some resistance but indicates that stability has been of shorter duration and of lower resistance to erosion during major flows. Lack of vegetation, or its recent burial by sediments, indicates that drainage-line erosion and sedimentation processes are active during runoff events.

Two indicators are scored when assessing vegetation within an ephemeral drainage line: vegetation on drainage-line side walls, and vegetation on drainage-line floors.

3. Ephemeral drainage-line wall vegetation indicator

This indicator assesses the amount of vegetation covering and protecting the walls of the ephemeral drainage line. Drainage-line walls with a dense covering of vegetation will resist the erosive forces of flows into and along the channel.

4. Ephemeral drainage-line floor vegetation indicator

This indicator evaluates the type and amount of vegetation on the floor, or bed, of the ephemeral drainage line using similar methods to those employed in the assessment of vegetation along channel walls. Drainage-line floors with little perennial vegetation, or with sparse ephemeral vegetation, provide scant protection from the forces of flows into and down the drainage line.

When assessing the above two indicators, no distinction is made between native and exotic vegetation because this assessment is aimed at defining the degree of protection of ephemeral drainage lines by vegetation. However, the presence of exotic weeds within a drainage line has implications for RPs who may need to control these weeds with chemicals or by physically removing them without reactivating erosion.

Assessing Indicators of Ephemeral Drainage-line Shape

The shape of an ephemeral drainage line strongly affects the concentration, or dispersion, of energy in the water flowing into and down the drainage line. This in turn affects the extent to which materials are eroded and deposited along the drainage channel. The shape of a channel and its relationship with the wider floodplain also affects floodplain-forming processes. These geomorphic processes are complex and beyond the scope of this book. We refer those interested in these processes to Brierley and Fryirs (2005).

Two indicators are used to assess the vigor of these processes at selected points along the gully: cross-section shape, and longitudinal profile. The types of sediment deposits and erosion features associated with different drainage-line shapes and profiles help RPs to score these two indicators.

5. Ephemeral drainage-line cross-section indicator

Ephemeral drainage lines that are still actively eroding typically have channel banks with steep or vertical walls and sharp edges, whereas those that are stabilizing tend to have wall angles of less than 65° and more rounded edges. Thus the cross-sectional shape of a drainage line gives a strong indication of whether it is stabilizing or continuing to erode and produce sediment.

6. Ephemeral drainage-line longitudinal-section indicator

The longitudinal profile of a drainage line indicates the pattern and strength of flows along the channel. When evaluating this profile it is important for RPs to be aware of changes in slope. For example, the bed may be observed to have alternating flat and sloping sections, which indicates differences in flow energy. The longitudinal profile also indicates how the drainage line interacts with the adjacent floodplain by characterizing a continuum that ranges from deeply incised gullies to a chain-of-ponds profile. (See figure 15.5.) In a chain-of-ponds profile the drainage line and floodplain are connected, and floodwaters dissipate with low energy onto the bordering floodplain. In deeply incised drainage-line profiles the channel bed and floodplain are disconnected and floodwaters further incise the channel instead of dissipating over the floodplain.

Figure 15.5. A naturally stabilized channel characterized as a "chain of ponds" with low flow energy.

Assessing Indicators of Ephemeral Drainage-line Erodability

Many Australian soils slake and disperse when wet. These soils readily erode when exposed to flows of water. Tunnel erosion may occur, even at low-flow velocities, when dispersive soils become wet, which may appear as sink holes on a hillslope. Other signs of erosion include fluting, undercutting, and caving along incised drainage-line walls, and mass wasting onto floors. These erosion features have many implications for RPs aiming to restore landscapes. RPs need to be asking if the erosion process is predominantly due to initial exposure of dispersive materials that have then continued to erode when exposed to low- and high-flow events. Also, how can this erosion be reduced?

To answer such questions, RPs need to evaluate the stability of soils found along drainage lines before assessing indicators. The slake test is a simple but useful way to evaluate the degree to which soil particles disintegrate into smaller particles that settle and readily erode. (See chapter 14.) Another test, the aggregate stability in water test (ASWAT), evaluates whether clay dispersion is occurring. Clay dispersion is a much more serious erosion threat than particles that just disintegrate and settle (slake). RPs will find that restoring drainage lines with exposed dispersive clays is technically more difficult and expensive than dealing with soil-particle slumping (slaking).

Like the slake test, the ASWAT looks at the response of dry soil fragments when immersed in high-quality water, but scores the degree of milkiness (amount of clay dispersion) after ten minutes and then again after two hours. Dispersion indices from zero (no dispersion) to 16 (complete dispersion) emerge, which are based on the rapidity and degree of dispersion. Typically, ASWAT index values of 6 or more would signify to RPs a problem urgently needing attention. Theoretical details of the ASWAT soil stability test are available in Field et al. (1997). RPs will find instructions on how to perform the ASWAT test in an LFA document available online (http://members.iinet.net.au/~lfa_procedures/). RPs can conduct both the ASWAT and slake tests in the field or, if a large number of samples need assessment, in the laboratory.

After evaluating the stability of drainage-line wall and floor soil materials, RPs can assess two indicators: drainage-line wall erodability, and drainage-line floor erodability.

7. Ephemeral drainage-line wall erodability indicator

This indicator reflects the susceptibility of ephemeral drainage-line walls to erosion by runoff flows, both by overland flows spilling over walls and by energetic flows along channel floors. RPs score this indicator based on the degree of exposure of unstable soil materials along the drainage-line wall.

8. Ephemeral drainage-line floor erodability indicator

The size and cohesion of the materials on the floor of an ephemeral drainage line indicates its potential erodability. RPs use these attributes to score this indicator. For example, large rocks protect drainage-line floors from erosion by dispersing the energy of flows. Floor materials that are loose and similar in size, or smaller than those on drainage-line walls, require less energetic flows to mobilize them and, hence, indicate to RPs that these drainage-line floors are susceptible to erosion. The one exception to this is the organic matter/clay material found within a chain of ponds. (See figure 15.5.) The floors of these ponds are relatively stable because of the very low flow downstream energy in such locations.

Recording and Using Ephemeral Drainage-line Assessments

Starting at the top of an ephemeral drainage line, we recommend that RPs first move down the drainage line and stratify it into zones reflecting different states of health or stability along the line. Then at the mid-point of each identified zone, RPs can efficiently assess and score the eight indicators described above using the methods described in the EDA document available online (http://members.iinet.net.au/~lfa_procedures/). These data can then be used to produce a map reflecting differences in the stability along the ephemeral drainage line. These data can also be used to estimate an average index of overall drainage-line health. However, we have found that it is more useful to map drainage-line stability in sections or zones rather than trying to get an overall average for the whole drainage line. There can be quite surprising changes, from very stable to very unstable, and the reverse, along ephemeral drainage lines. We have found that it cannot be assumed that a lack of stability in upper reaches of a drainage line will necessarily transfer right down the drainage line. A number of differing restoration procedures may need to be designed and applied on different drainage-line locations to properly address various problems along a drainage line. (See, for example, Lane 2008.)

The RP can classify the underlying causes of erosion activity in different sections of an ephemeral drainage line by looking at the eight indicator scores for that section. This will show, for example, whether an actively eroding section is caused by high flow rates from adjoining slopes, or exposure of unstable (dispersive) side-wall material to flows. By assessing ephemeral drainage-line stability in sections, RPs can determine where the most urgent remedial actions are needed, and which technologies to apply.

Chapter 16

Vegetation Assessments: Structure and Habitat Complexity Indicators

Back in chapter 2 we illustrated the concept that the physical presence, spatial pattern, and structure of vegetation in a landscape system serve important roles in a number of critical positive feedback processes. For example, pulses of plant growth can add to the density and size (width and height) of vegetation patches, which then function to enhance the capacity of the landscape to capture rainwater and dust particles in future storm events. We call this the *functional role* of vegetation. Patches of vegetation having a complex mix of plant species and life-forms also provide habitats for animals and the goods and services needed by humans.

Here in chapter 16 we describe a number of attributes of vegetation structure that restoration practitioners (RPs) can measure along gradsects, including vegetation density and size, composition, and vertical cover. We also describe how horizontal and vertical vegetation measurements and other site factors can be used to estimate an index of habitat complexity.

Assessing Vegetation Density and Size

As noted, the density and the size of vegetation patches (whether measured as individual plants or as groupings of plants) are very important in determining how landscapes function. RPs can simply estimate vegetation-patch density and size by counting the number of patches, and measuring their lengths and widths, along a landscape func-

tion analysis (LFA) gradsect positioned in the landscape. (See chapter 13.) We have found that a simple linear expression of density (i.e., number of patches per length of line transect) is sufficient for most purposes of monitoring restoration sites. However, RPs may find that some stakeholders prefer a report using area-based vegetation density, which is the number of individuals or patches per unit area. Area-based density can be estimated along LFA gradsects by making additional measurements using plot-based or plotless-based vegetation sampling methods

Plot-based Density

With this method, RPs establish plots of known area (e.g., 1 m^2), which are positioned either regularly (e.g., every 20 m) or randomly along the gradsect. Then they simply count individual plants (or vegetation patches) within each plot. By counting a number of plots, which represents a sample for the site, the RP can calculate the average number of individuals (or patches) per unit area (e.g., mean number per square meter). If a site has high spatial heterogeneity, the RP needs to count a large number of plots to adequately sample the site, which can be very time consuming and tedious. The number of plots needed to adequately sample a site can be estimated by using equations in an LFA document on sample size available online (http://members .iinet.net.au/~lfa_procedures/).

Plotless-based Density

We use a plotless method based on measuring distances from a point to individual plants (or to the center of patches), or from individual to individual, which are called *point-centered quarter* (PCQ) and *wandering quarter* (WQ) methods, respectively. (See figure 16.1.) For example, we measure PCQ distances at points positioned regularly along gradsects (e.g., every 20 m). These distances are used to calculate mean area, which is the average area occupied by an individual plant or patch (e.g., 10 m² per individual). Density is simply the reciprocal of mean area (e.g., 1/10 = 0.1 individuals per m²). These and other plotless methods are described by Bonham (1989). To tabulate PCQ and WQ data, and for computing mean area and density, RPs can use an LFA vegetation data entry spreadsheet available online (http://members.iinet.net.au/~lfa_procedures/).

To estimate the density of trees and shrubs, we have found plotless methods particularly useful because they are typically much more time efficient than plot-based counting procedures. This is especially true if large plots need to be established

and surveyed to obtain enough counts for an adequate sample. Quinn and Keough (2002) discuss adequacy of sampling, which is an important issue when estimating density of vegetation.

Plotless procedures have been widely used in forestry not only to estimate tree densities but also to estimate tree size. When the distance is measured from a point to a tree, the diameter of the tree can also be measured (e.g., diameter at breast height [DBH], or more consistently at 1.3 m). These distance and diameter measurements can then be used to estimate tree basal area, which is typically expressed as square meters per hectare. This estimate of tree size is a very useful measure for indicating how trees are developing on sites being rehabilitated to savanna, woodland, or forest landscapes.

Assessing Vegetation Composition

When using plotless methods to measure distances to, and diameters of, trees, we have found it useful to identify the species of each tree being measured. This is because changes in tree species composition

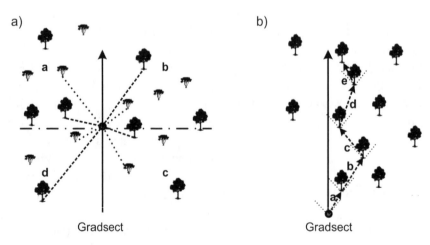

FIGURE 16.1. Plotless vegetation sampling procedures: (a) point-centered quarter method, where distances from a number of points selected along a gradsect are measured to trees (dashed-lines) and shrubs (dotted lines) in each of four quarters (a–d). (b) Wandering quarter method, where distances from a starting point selected on a gradsect are measured between trees (dashed-lines, a–e) within a quarter (90° angle) opening in the direction of the gradsect. Wandering quarter distances can also be measured between shrubs or grasses (for brevity, not shown).

can be very important for indicating progress of rehabilitated sites toward the state expected from the composition of reference sites. For example, in chapter 4, we described a case study where reference sites located in the natural savannas in northern Australia are typically dominated by slow-growing *Eucalyptus* species, whereas rehabilitated sites are often initially dominated by fast-growing, fire-prone *Acacia* species. Ideally, acacias should decline with time as eucalypts develop on rehabilitated sites in savannas. If this trend does not occur, it indicates that the site may be trapped in a fire cycle in which acacias continue to dominate (by regenerating from soil seed banks) while small eucalypts are killed by repeated fires before they can grow large enough to become fire resistant and set seed.

When making LFA measurements along gradsects (e.g., by a plotless method), we have also found it useful to record the identity of shrub and perennial grass species. These data are important, for example, because the presence of fast-growing exotic shrubs or weeds may indicate a rehabilitation trend that is not toward a desired goal. Although exotic grasses are often sown on rehabilitation sites, in many cases the goal is to reestablish natural grasslands and savannas, hence, the presence and increasing abundance of exotic grasses on a rehabilitated site is a negative indicator. In addition, because gradsect data are spatially referenced, differences in density or species composition can be tracked across the landscape.

Assessing Vegetation Vertical Structure

When measuring vegetation patch size, density, and species composition within life-forms (e.g., trees, shrubs, perennial grasses) using plot or plotless methods, we have found it useful and convenient to measure other attributes at the same time. For example, the architecture of tree canopies can be defined by measuring overall tree height, the height from the ground to the bottom of the foliage, and the width and breadth of the canopy. (See figure 16.2.) These

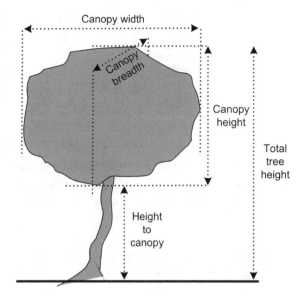

FIGURE 16.2. A diagram illustrating the different canopy structural attributes that can be measured on a tree.

features could also be measured on other life-form categories such as tall shrubs, low shrubs, and the ground vegetation.

If such data by life-form groups are obtained by the plotless method and recorded in the LFA vegetation data entry spreadsheet, the RP will find an embedded set of equations that calculate the number of plants per hectare and the canopy cover (square meters per hectare) and volume (cubic meters per hectare) for each plant life-form sampled. This spreadsheet also has an option for plotting the vertical distribution of canopy cover (square meters per hectare) at height increments of 1 m. For example, the vertical cover frequency distribution of combined grass, shrub, and tree canopies on a site can indicate how the landscape on a rehabilitation site is developing toward that expected from a reference site. (See figure 16.3.)

Such vertical profiles of vegetation structure provide RPs with useful functional interpretations. For example, we have observed woodland landscapes where a reference site had a well-developed shrub canopy layer near the ground. (See figure

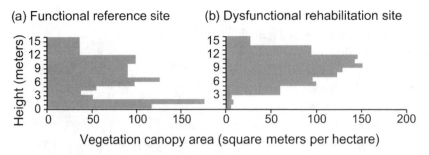

FIGURE 16.3. (a) The vertical distribution of woodland vegetation canopy areas plotted here at 1 m height intervals on a highly functional reference site. (b) A dysfunctional rehabilitation site where the ground and shrub layers, up to about 3 m, have been eliminated by grazing animals.

16.4a.) On this site the shrub canopy merges with the tree canopy layer to provide a continuous vertical canopy cover. In contrast, we observed a nearby disturbed site that lacked both a shrub canopy and grass cover. (See figure 16.4b.) This lack of lower vegetation was caused by excessive livestock grazing and trampling. This means that this site is partially dysfunctional and needs active rehabilitation because of the lack of vegetation on, or just above, the surface. These surface and low vegetation layers would normally protect the landscape from excessive erosion by water and wind.

Because data from reference sites are used to establish expected values, the monitoring and analysis of vegetation vertical structure data over time is essentially equivalent to the monitoring and analysis of soil-surface condition indices. (See chapter 14.) In both cases, restoration site indicator values should trend toward values expected from reference sites.

Assessing Habitat Complexity

Another reason for making additional measurements along gradsect lines, such as measures on tree canopy architecture, is that these measurements can also be used by RPs to estimate an index known as *habitat complexity*. This index assesses the extent to which a site is functioning to provide habitats (shelter, food, nesting sites) required by vertebrate fauna. The habitat complexity index was developed to monitor and assess habitat availability for small to medium-sized tree-dwelling mammals in the forests of southeast Australia; these forests were being affected by logging. (See Coops and Catling 1997.)

In their forest study, Coops and Catling based their habitat complexity index on observing, measuring, and classifying a number of site features, which included the following:

- tree canopy cover (percentage)
- shrub canopy cover (percentage)
- ground vegetation density (two classes: sparse or dense)
- ground vegetation height (two classes: less than 0.5 m or greater than 0.5 m)
- ground covered by rocks, fallen logs and other surface debris (percentage)
- general amounts of soil moisture (two classes: dry or moist)
- availability of free water (two classes: free water nearby or site wet and waterlogged)

They combined these seven features into five attributes, which were scored as being in one of four classes from 0 to 3. (See table 16.1.) The habitat complexity index was calculated as the sum of five assessed scores, which ranged from 0 (if all landscape features were scored as 0) to 15 (if features were all scored as 3).

(a)   (b)

FIGURE 16.4. (a) A highly functional woodland has a complex vertical distribution of vegetation from the ground layer up to the top of the tree canopy. (b) A woodland heavily disturbed by livestock lacks vegetation structure at heights below 3 m.

For their forest sites, Coops and Catling found that habitat complexity index values at or near 15 would be obtained only in forests with closed or nearly closed canopies (tree and shrub canopy greater than 70 percent); with dense, tall, ground vegetation (greater than 0.5 m); with abundant ground debris (greater than 70 percent); and with permanently available water. (See table 16.1.) Hence, in woodlands, savannas, shrublands, and grasslands, using these features and this scoring system, habitat complexity index values would always total less than 15.

Thus, the habitat complexity index is the only method in the LFA set of procedures that might benefit from redesign so that it more usefully applies to nonforest vegetation types and indicates the specific habitat requirements of faunal groups other than tree-dwelling mammals. All the previous LFA procedures can be used in any biome without modification, because in every case, a process, rather than a specific biological entity, is being evaluated.

Although more appropriate site features or indicators could be devised by RPs when applying the

TABLE 16.1

Site features are scored from 0 to 3 to calculate the habitat complexity index

Feature	Score			
	0	1	2	3
Tree canopy (%)	0	<30	30–70	>70
Shrub canopy (%)	0	<30	30–70	>70
Ground herbage (density and height)	Sparse <0.5 m	Sparse > 0.5 m	Dense <0.5 m	Dense > 0.5 m
Debris (logs, rocks) (%)	0	<30	30–70	>70
General soil moisture and water availability	Dry	Moist	Permanent water nearby	Wet

habitat complexity index to nonforest landscapes, scientifically verifying such index modifications would be time consuming and expensive. Because habitat complexity is a general index of habitat quality, rather than the presence of particular animals, it is often easier and less costly for RPs to directly survey the presence and abundance of the fauna of interest than to modify the habitat complexity index.

Chapter 17

Reflections on Restoring Landscapes:
A Function-Based Adaptive Approach

In this final chapter we reflect on the function-based adaptive approach we recommend for restoring landscapes. Our case studies and scenarios illustrated how, if restoration practitioners (RPs) adhere to a series of logical steps and principles, they will achieve positive trends toward successfully rehabilitating disturbed landscapes. Importantly, if RPs find negative trends, the adaptive approach leads them to apply timely adjustments or corrections to restoration technologies. Returning briefly to the analogy of planning a journey using a road map, knowing one's starting point and destination are very important, but choosing the best route relies on frequently gathering updates on progress (monitoring) so that decisions on any detours en route are based on the most up-to-date information available.

Applying this to landscape restoration, progress can be conceptualized as a series of constructive stages toward achieving restoration goals. (See figure 17.1.) Each stage builds on earlier stages. The ultimate goal is to build a complex landscape that possesses a multiplicity of life-forms (biological diversity) and regulatory processes (functional diversity). Such landscapes will be buffered against environmental and management disturbances both by their accumulated natural capital and by the

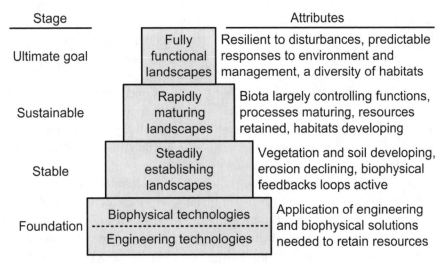

Stage		Attributes
Ultimate goal	Fully functional landscapes	Resilient to disturbances, predictable responses to environment and management, a diversity of habitats
Sustainable	Rapidly maturing landscapes	Biota largely controlling functions, processes maturing, resources retained, habitats developing
Stable	Steadily establishing landscapes	Vegetation and soil developing, erosion declining, biophysical feedbacks loops active
Foundation	Biophysical technologies ---- Engineering technologies	Application of engineering and biophysical solutions needed to retain resources

FIGURE 17.1. A four-tier pyramid illustrating how goals and attributes for rehabilitating landscapes are achieved in stages. Adapted from figure 9.5 in Tongway and Ludwig (2007).

complex diversity of physical and biological processes responsible for new natural capital accession. See Aronson et al. (2007) for a discussion of natural capital as a concept and process. The key test for natural capital assessment is how well landscape systems gain resources by retention, utilization, and cycling processes. Finally, social acceptance of the rehabilitated landscape is part of the ultimate evaluation of whether restoration goals have been achieved.

We intentionally selected the shape of figure 17.1 to emphasize the concept that progression from one phase of restoration to another requires that the first phase be fully realized to provide a solid functional base on which to build succeeding phases. We also intentionally divided the attributes of the first foundation stage into two components because engineering and biophysical technologies are both required to build the foundation for restoring a disturbed site. However, the emphasis on the use of engineering versus biophysical technologies will differ between applications. For example, mine-site reclamation typically requires designing and constructing new landforms (physical technologies), whereas renewing damaged rangelands, where the original landforms and soils are still largely intact, mostly requires reestablishing perennial vegetation (biological technologies). Figure 3.1 in chapter 3 illustrates how physical and biological processes initiated by these technologies develop over time as landscapes become rehabilitated.

The information needed to show whether each stage in figure 17.1 has been reached is obtained by monitoring the indicators described in chapters 13 to 16. We specifically selected these indicators because they assess the functional attributes listed as examples at each stage in the stepped pyramid. Monitoring these indicators is especially important in the early, vulnerable stages of restoration.

Although we show sharp boundaries between stepped stages in figure 17.1, there will of course be a gradual transition between progressive stages similar to the continuous progress of a restoration indicator along a continuum of landscape functionality, as illustrated in figure 3.6 in chapter 3. We

feel that it is the role of indicator continuums, and the pyramid in figure 17.1, to invite RPs to use monitoring data to reflect on their restoration progress.

Our five-step adaptive approach provides the action steps that RPs take to successfully restore disturbed landscapes. (See figure 1.1.) In figure 17.1, our stepped pyramid aims to reflect on the accumulation or the building of knowledge that RPs gain to assess progress through each landscape restoration stage. For example, in a rehabilitating forest the pyramid may show RPs that their ultimate goal has not yet been reached because time has been insufficient for trees to develop hollows as nesting sites; however, their measurements may indicate significant progress toward this goal. Moreover, as stages are reached they can be descriptively and functionally defined and, if necessary, defended as, for example, in cases of required closure criteria in mine-site rehabilitation.

We have noticed in our professional lives that monitoring indicators of landscape restoration is frequently relegated to junior staff or treated as an unwelcome necessity—it is often done only because it is a legal requirement. We contend, however, that monitoring is an important and valuable task essential and integral to landscape restoration. Monitoring is designed for RPs to inform stakeholders about restoration trends and thus establish priorities for ongoing project activities. Early warnings that trends are not OK can lead to technology adjustments and corrections before repairs become costly. Well-designed monitoring provides RPs with the data they require to assess and evaluate the effectiveness of their treatments as early as possible, which avoids wasted time, effort, and expense in the future.

Methods for monitoring landscape restoration indicators are described in documents available to RPs online (http://members.iinet.net.au/~lfa_procedures/). These methods represent a toolbox from which RPs can select specific tools to measure the information they need to promptly detect and fix problems in the earliest stages of landscape restoration. The tools (monitoring methods) used will improve and change over time (and online documents will be updated) as we all learn more about the biophysical processes and engineering solutions involved in restoring the functionality of landscapes. We all know that delaying the commencement of monitoring risks missing opportunities to improve restoration practices. If "justice delayed is justice denied," then monitoring delayed is opportunity missed.

For some RPs, switching from simply observing the presence, absence, or abundance of organisms to assessing the status of functional processes in an explicitly spatial (landscape) context is as challenging as venturing into previously uncharted waters. We are confident that RPs who take up this challenge will find that it is well worth the effort, because the skills they acquire are readily transferrable to a multitude of different landscapes. Being able to "read the landscape" is a rare and valuable asset.

References

Aronson, J., S. J. Milton, and J. N. Blignaut, eds. 2007. *Restoring natural capital: Science, business, and practice.* Washington, DC: Island Press.

Bastin, G. 1991. Rangeland reclamation on Atartinga Station, Central Australia. *Australian Journal of Soil and Water Conservation* 4:18–25.

Bonham, C. D. 1989. *Measurements for terrestrial vegetation.* New York: John Wiley and Sons.

Brierley, G. J., and K. A. Fryirs. 2005. *Geomorphology and river management: Applications of the River Styles Framework.* Oxford, UK: Blackwell Publishing.

Coops, N. C., and P. C. Catling. 1997. Utilizing airborne multispectral videography to predict habitat complexity in eucalypt forests for wildlife management. *Wildlife Research* 24:691–703.

Cunningham, G. M. 1987. Reclamation of scalded land in western New South Wales: A review. *Journal of Soil Conservation NSW* 43:52–61.

Daily, G. C., ed. 1997. *Nature's services: Societal dependence on natural ecosystems.* Washington, DC: Island Press.

Du Toit, J. T., K. H. Rogers, and H. C. Biggs, 2003. *The Kruger experience: Ecology and management of savanna heterogeneity.* Washington, DC: Island Press.

Dye, P. J., C. Jarmain, B. Oageng, J. Xaba, and I. M. Weiersbye. 2008. The potential of woodlands and reed-beds for control of acid mine drainage in the Witwatersrand gold fields, South Africa. In *Proceedings, Third International Mine Closure Seminar, Johannesburg, South Africa,* ed. A. Fourie, M. Tibbett, I. Weiersbye, and P. Dye, 487–97. Perth, Australia: Australian Centre for Geomechanics.

Eldridge, D. J., and R. S. B. Greene. 1994. Microbiotic soil crusts: A review of their roles in soil and ecological processes in the rangelands of Australia. *Australian Journal of Soil Research* 32:389–415.

Ellis, T. W., S. Leguédois, P. B. Hairsine, and D. J. Tongway. 2006. Capture of overland flow by a tree belt on a pastured hillslope in south-eastern Australia. *Australian Journal of Soil Research* 44:117–25.

Field, D. J., D. C. McKenzie, and A. J. Koppi. 1997. Development of an improved Vertisol stability test for SOILpak. *Australian Journal of Soil Research* 35:843–52.

Friedel, M. H., J. E. Kinloch, and W. J. Muller. 1996. The potential of some mechanical treatments for rehabilitating arid rangelands. 1. Within-site effects and economic returns. *Rangelands Journal* 18:150–64.

Friedel, M. H., H. Puckey, C. O'Malley, M. Waycott, A. Smyth, and G. Miller. 2006. *Buffel grass: Both friend and foe. An evaluation of the advantages and disadvantages of buffel grass use and recommendations for future research.* Research Report 17. Alice Springs, Australia: Desert Knowledge Cooperative Research Centre.

Gillison, A. N., and K. R. Brewer. 1985. The use of gradient directed transects or gradsects in natural resource surveys. *Journal of Environmental Management* 20:103–27.

Giurgevich, B. 1999. Reclamation success standards for coal surface mine lands in the western United States. In *People of the rangelands: Building the future. Proceedings of the 6th International Rangeland Congress*, ed. D. Eldridge and D. Freudenberger, 957–61. Adelaide, Australia: Australian Rangeland Society.

Gould, S. 2010. Does post-mining rehabilitation on the Weipa bauxite plateau restore bird habitat values? PhD diss., Australian National University, Canberra, Australia.

Haagner, A. S. H. 2009. The role of vegetation in characterising landscape function on rehabilitating gold tailings. MS thesis, North-West University, Potchefstroom, South Africa.

Hollingsworth, I. 2010. Mine landform design using natural analogues. PhD diss., University of Sydney, Sydney, Australia.

James, C. D., J. Landsberg, and S. R. Morton. 1999. Provision of watering points in the Australian arid zone: A review of effects on biota. *Journal of Arid Environments* 41:87–121.

Kellner, K., and A. S. Moussa. 2009. A conceptual tool for improving rangeland management decision-making at grassroots level: The local-level monitoring approach. *African Journal of Range and Forage Science* 26:139–48.

Lacy, H. W. B., and K. L. Barnes. 2006. Tailings storage facilities: Decommissioning planning is vital for successful closure. In *Proceedings, First International Seminar on Mine Closure*, ed. A. Fourie and M. Tibbett, 139–48. Perth, Australia: Australian Centre for Geomechanics.

Landsberg, J., C. D. James, S. R. Morton, W. Muller, and J. Stol. 2003. Abundance and composition of plant species along grazing gradients in Australian rangelands. *Journal of Applied Ecology* 40:1008–24.

Lane, S. 2008. Stream health and sediment sources of Pierces Creek Catchment. MS thesis, Australian National University, Canberra, Australia.

Leguédois, S., T. W. Ellis, P. B. Hairsine, and D. J. Tongway. 2008. Sediment trapping by a tree belt: Processes and consequences for sediment delivery. *Hydrological Processes* 22:2523-34.

Loch, R. 1997. Landform design: Better outcomes and reduced costs by applying science to above and below ground issues. In *Proceedings, 22nd Annual National Environmental Workshop*, 550–63. Adelaide, Australia: Minerals Council of Australia.

Loch, R., T. Stevens, G. Wells, and R. Gerrard. 2006. Development of key performance indicators for rehabilitation, Murrin Murrin Nickel Operation. In *Proceedings, First International Seminar on Mine Closure*, ed. A. Fourie and M. Tibbett, 569–76. Perth, Australia: Australian Centre for Geomechanics.

Lovett, S., J. Lambert, J. E. Williams, and P. Price. 2008. *Restoring landscapes with confidence: An evaluation of the science, the methods and their on-ground application.* Final Report. Canberra, Australia: Land and Water Australia.

Ludwig, J. A., and J. A. Courtenay. 2008. *Savannah Way landscapes.* Darwin, Australia: Tropical Savannas Cooperative Research Centre, Charles Darwin University.

Ludwig, J. A., N. Hindley, and G. Barnett. 2003. Indicators for monitoring mine site rehabilitation: Trends on waste-rock dumps, northern Australia. *Ecological Indicators* 3:143–53.

Ludwig, J. A., and D. J. Tongway. 1996. Rehabilitation of semi-arid landscapes in Australia. 2. Restoring vegetation patches. *Restoration Ecology* 4:398–406.

———. 2000. Viewing rangelands as landscape systems. In *Rangeland Desertification*, ed. O. Arnalds and S. Archer, 39–52. Dordrecht, The Netherlands: Kluwer Academic Publishers.

———. 2002. Clearing savannas for use as rangelands in Queensland: Altered landscapes and water-erosion processes. *Rangeland Journal* 24:83–95.

Ludwig, J. A., D. J. Tongway, D. A. Freudenberger, J. C. Noble, and K. C. Hodgkinson, eds. 1997. *Landscape ecology, function and management: Principles from Australia's rangelands.* Melbourne, Australia: CSIRO Publishing. (Out of print: available as pdf files online at http://members.iinet.net.au/~lfa_procedures/)

McKenzie, N. J., D. J. Jacquier, and A. J. Ringrose-Voase. 1994. A rapid method for estimating soil shrinkage. *Australian Journal of Soil Research.* 32:931–38.

Milton, S. J., W. R. J. Dean, and D. M. Richardson. 2003. Economic incentives for restoring natural capital: Trends in southern African Rangelands. *Frontiers in Ecology and the Environment* 1:247–54.

Munro, N. T., D. B. Lindenmayer, and J. Fischer. 2007. Faunal response to revegetation in agricultural areas of Australia: A review. *Ecological Management and Restoration* 8:199–207.

Noble, J. C. 1997. *The delicate and noxious scrub: CSIRO studies on native tree and shrub proliferation in the semi-arid woodlands of eastern Australia.* Canberra, Australia: CSIRO Sustainable Ecosystems.

Noble, J. C., A. C. Grice, M. J. Dobbie, W. J. Muller, and J. T. Wood. 2001. Integrated shrub management in semi-arid woodlands of eastern Australia: Effects of chemical defoliants applied after an initial disturbance. *Rangeland Journal* 23:224–58.

Oades, J. M. 1993. The role of biology in the formation, stabilization and degradation of soil structure. *Geoderma* 56:377–400.

Peters, T. H. 1984. *Rehabilitation of mine tailings: A case of complete ecosystem reconstruction and revegetation of industrially stressed lands in the Sudbury Area, Ontario, Canada.* New York: Wiley and Sons.

Purvis, J. R. 1986. Nurture the land: My philosophies of pastoral management in Central Australia. *Australian Rangeland Journal* 8:110–7.

Quinn, G. P., and M. J. Keough. 2002. *Experimental design and data analysis for biologists.* Cambridge, UK: Cambridge University Press.

Rethman, N. F. G., P. D. Tanner, M. E. Aken, and R. Garner. 1999. Assessing reclamation success of mined surfaces in South Africa with particular reference to strip coal mines in grassland areas. In *People of the rangelands: Building the future. Proceedings of the 6th International Rangeland Congress,* ed. D. Eldridge and D. Freudenberger, 949–53. Adelaide, Australia: Australian Rangeland Society.

Reynolds, J. F., and D. M. Stafford Smith, eds. 2002. *Global desertification: Do humans cause deserts?* Dahlem Workshop Report 88. Berlin, Germany: Dahlem University Press.

Richards, S., and D. Walsh. 2008. A reassessment of soil rehabilitation works established on Bond Springs Station (Alice Springs) in 1968. *Range Management Newsletter* 08/3:4–8.

Ringrose-Voase, A. J., D. W. Rhodes, and G. F. Hall. 1989. Reclamation of a scalded Red Duplex Soil by waterponding. *Australian Journal of Soil Research* 27:779–95.

Robin, L. 2007. *How a continent created a nation.* Sydney, Australia: University of New South Wales Press.

Ross, K. A., J. E. Taylor, M. D. Fox, and B. J. Fox. 2004. Interaction of multiple disturbances: Importance of disturbance interval in the effects of fire on rehabilitating mined areas. *Austral Ecology* 29:508–29.

Ryan, J. G., J. A. Ludwig, and C. A. McAlpine. 2007. Complex adaptive landscapes (CAL): A conceptual framework of multi-functional, non-linear ecohydrological feedback systems. *Ecological Complexity* 4:113–27.

Ryan, J. G., C. A. McAlpine, and J. A. Ludwig. 2010. Integrated vegetation designs for enhancing water retention and recycling in agroecosystems. *Landscape Ecology.* Published online: DOI 10.1007/s10980-010-9509-7.

Setyawan, D. 2004. Soil development, plant colonization and landscape function analysis for disturbed lands under natural and assisted rehabilitation. PhD diss., University of Western Australia, Perth, Australia.

Spain A. V., J. Esterle, and T. P. T. McLennan. 1995. Information from geology: Implications for soil formation and rehabilitation in the post coal mining environment, Bowen Basin, Australia. In *Proceedings of the Bowen Basin symposium—1995: 150 years on,* ed. I. L. Follington, J. W. Beeston, and L. H. Hamilton, 147–55. MacKay, Queensland, Australia: Coal Geology Group, Geological Society of Australia.

Spain, A. V., J. A. Ludwig, M. Tibbett, and D. J. Tongway. 2009. *Ecological and minesoil development studies at the Rio Tinto Gove Mine Site, Northern Territory.* Report, Perth, Australia: University of Western Australia Centre for Land Rehabilitation.

Thiry, M., and R. Simon-Coincon, eds. 1999. *Palaeoweathering, palaeosurfaces, and related continental deposits.* Special Publication 27 of the International Association of Sedimentologists Series. Oxford, UK: Blackwell Science.

Thomson, R. 2008. Waterponding: Reclamation technique for scalded duplex soils in western New South Wales rangelands. *Ecological Management and Restoration* 9:170–81.

Tongway, D. J. 1995. Monitoring soil productive potential. *Environmental Monitoring and Assessment* 37:303–18.

Tongway, D. J., and N. Hindley. 2000. Assessing and monitoring desertification with soil indicators. In *Rangeland desertification*, ed. O. Arnalds and S. Archer, 89–99. Dordrecht, The Netherlands: Kluwer Academic.

———. 2004. *Landscape function analysis: Procedures for monitoring and assessing landscapes with special reference to minesites and rangelands.* Canberra, ACT, Australia: Sustainable Ecosystems, Commonwealth Scientific and Industrial Research Organisation. (Updated as an "LFA Field Procedures," which is available online: http://members.iinet.net.au/~lfa_procedures/).

Tongway, D. J., N. Hindley, J. A. Ludwig, A. Kearns, and G. Barnett. 1997. Early indicators of ecosystem rehabilitation on selected minesites. In *Proceedings of the 22nd Annual Environmental Workshop*, 494–504. Dickson, ACT, Australia: Minerals Council of Australia.

Tongway, D. J., and J. A. Ludwig. 1996. Rehabilitation of semi-arid landscapes in Australia. 1. Restoring productive soil patches. *Restoration Ecology* 4:388–97.

———. 1997. The nature of landscape dysfunction in rangelands. In *Landscape ecology, function and management: Principles from Australia's rangelands*, ed. J. A. Ludwig, D. J. Tongway, D. A. Freudenberger, J. C. Noble, and K. C. Hodgkinson, 49–61. Melbourne, Australia: CSIRO Publishing.

———. 2001. Theories on the origins, maintenance, dynamics and functioning of banded landscapes. In *Banded vegetation patterning in arid and semiarid environments: Ecological processes and consequences for management*, ed. D. J. Tongway, C. Valentin, and J. Segheri, 20–31. New York: Springer Science.

———. 2002a. Australian semiarid lands and savannas. In *Handbook of ecological restoration.* Vol. 2. *Restoration in practice*, ed. M. R. Perrow and A. J. Davy, 486–502. Cambridge, UK: Cambridge University Press.

———. 2002b. Desertification, reversing. In *Encyclopedia of soil science*, ed. R. Lal, 343–45. New York: Marcel Dekker.

———. 2005. Heterogeneity in arid and semiarid lands. In *Ecosystem function in heterogeneous landscapes*, ed. G. M. Lovett, C. G. Jones, M. G. Turner, and K. C. Weathers, 189–205. New York: Springer Science.

———. 2006. Assessment of landscape function as an information source for mine closure. In *Proceedings, First International Seminar on Mine Closure*, ed. A. Fourie and M. Tibbett, 21–29. Perth, Australia: Australian Centre for Geomechanics.

———. 2007. Landscape function as a target for restoring natural capital in semiarid Australia. In *Restoring natural capital: Science, business, and practice*, ed. J. Aronson, S. J. Milton, and J. N. Blignaut, 76–84. Washington, DC: Island Press.

———. 2009. Landscape dynamics. In *The Princeton guide to ecology*, ed. S. Levin, 425–30. Princeton, New Jersey: Princeton University Press.

Tongway, D. J., A. D. Sparrow, and M. H. Friedel. 2003. Degradation and recovery processes in arid grazing lands of central Australia. Part 1: Soil and land resources. *Journal of Arid Environments* 56:301–26.

Van den Berg, L., and K. Kellner. 2005. Restoring degraded patches in a semi-arid rangeland of South Africa. *Journal of Arid Environments* 61:497–511.

Vasey, A., R. Loch, and G. Willgoose. 2000. Rough and rocky landforms. In *Proceedings 25th National Environmental Workshop*, 244–59. Perth, Australia: Minerals Council of Australia.

Wedd, R. F. 2002. Vegetation compositional and structural changes over 27 years on the revegetated minesite, Nhulunbuy, N.T. MS thesis, Charles Darwin University, Darwin, Australia.

Weiersbye, I. M. 2007. Global review and cost comparison of conventional and phyto-technologies for mine closure. In *Proceedings, Second International Mine Closure Seminar,* ed. A. Fourie, M. Tibbett, and J. Wiertz, 13–29. Perth, Australia: Australian Centre for Geomechanics.

Weiersbye, I. M., and E. T. F. Witkowski. 2007. Impacts of acid mine drainage on the regeneration potential of Highveld phreatophytes. In *Multiple use management of natural forests and woodlands: Policy refinements and scientific progress,* ed. J. J. Bester, A. H. W. Seydack, V. T. Vorster, K. J. Van der Merwe, and S. Dzivhani, 224–37. Pretoria, South Africa: Government of South Africa Department of Water Affairs and Forestry.

Weiersbye, I. M., E. T. F. Witkowski, and M. Reichardt. 2006. Floristic composition of gold and uranium tailings dams, and adjacent polluted areas, on South Africa's deep-level mines. *Bothalia* 36:101–27.

Westoby, M., B. Walker, and I. Noy-Meir. 1989. Opportunistic management for rangelands not at equilibrium. *Journal of Range Management* 42:266–74.

Whisenant, S. G. 1990. *Repairing damaged wildlands: A process-orientated, landscape-scale approach.* Cambridge, UK: Cambridge University Press.

Wiedemann, H. T., and P. J. Kelly, 2001. Turpentine (*Eremophilla sturtii*) control by mechanical uprooting. *Rangeland Journal* 23:173–81.

Willgoose, G., and S. Riley, 1998. The long-term stability of engineered landforms of the Ranger Uranium Mine, Northern Territory, Australia: Application of a catchment evolution model. *Earth Surface Processes and Landforms* 23:237–59.

Williams, D. J., and J. T. Kline. 2006. Innovative mine closure design based on observations of mine and natural analogues. In *Proceedings, First International Seminar on Mine Closure,* ed. A. Fourie and M. Tibbett, 559–68. Perth, Australia: Australian Centre for Geomechanics.

Glossary

In defining the following terms used in this book we aimed to place them within the context of landscape restoration by building on terms defined in *The SER International Primer on Ecological Restoration* and relevant books on ecological restoration (see Further Reading).

abiotic. Nonliving or physical entities and processes such as rocks, rainfall, and wind.

adaptive approach. Making adjustments and corrections in restoration technologies in light of new information.

alien species. Fungi, plants, animals, and other organisms that have been introduced into a landscape in which they do not naturally occur.

alluvium. Sediments or other materials that have been transported by water and deposited within or outside a defined landscape area; these sediment deposits may be small to large.

anthropogenic. Caused primarily by the activities of people.

analogue. An area of landscape selected because its structure and functionality represent the desired goal for the landscape being restored; see also *reference site*.

ASWAT (aggregate stability in water test). The degree of "milkiness" observed if a natural, air-dry soil fragment (aggregate) disperses when immersed in water; stable soil fragments do not show signs of dispersion when wet; see also *slake test*.

bauxite. Mineral deposits rich in aluminium and iron.

biota. All biological species such as plants, animals, fungi, and microorganisms that occur at a given location.

biodiversity. The variety of living organisms at all levels of organization, including the genetic level, diversity within species, between species, and in ecosystems and landscapes; see also *functional diversity*.

biophysical processes. A combination of biological and abiotic factors that function together to moderate the way basic resources are retained and used in a landscape.

biopores. Channels or tubular openings through the soil left by the decay of plant roots and created by burrowing organisms such as termites. Biopores enhance the infiltration of water into soils.

brush packs. Piles of woody branches and twigs positioned along contours on sites being rehabiltated to slow flows of water and wind over surfaces, trapping soil sediments, litter, and seeds, and enhances infiltration.

cation exchange capacity. The capacity of the soil clay and organic matter complexes to adsorb and release cations in a solution.

community. An assemblage of organisms occurring in a landscape; typically used in combination with a taxonomic group (plant community, insect community, epiphyte community).

community structure. The physical appearance of a community as determined by the size, life-form, abundance, and spatial distribution of plant species.

degradation. A decline in landscape functionality and self-sustainability caused by stress and disturbance processes.

dispersivity. A property of soils defining how readily soil particles or aggregates disintegrate when wet; see also *ASWAT* and *slake test.*

disturbance. Natural or anthropogenic events such as cyclones and agriculture that change the structure, composition, and functioning of a landscape, often in a substantial way.

dysfunctional. A landscape that is no longer self-sustaining due to a breakdown of essential biophysical or socioeconomic processes.

ecosystem. A community of organisms interacting with one another and with the physical environment in the specified area where they live.

ecosystem goods and services. Materials such as food, fiber, and wood, and services such as the provision of clean drinking water, which contribute to the fulfilment of human needs.

ecological restoration. The process of assisting the recovery of ecosystems and landscapes that have been disturbed or damaged.

fragmentation. The division of a formerly continuous or homogeneous natural landscape into smaller units that are isolated from each other by other natural or anthropogenic land units.

framework species. A local species that, from its pioneer capacity, growth rate, canopy cover, modification of microclimate, and longevity provides ecological niches for many other species; a primary provider of ecosystem goods and services.

functional. A landscape that effectively self-regulates and utilizes available resources such as water and energy, and provides the goods and services required by populations, including humans, living in the landscape.

functional diversity. The variety of organisms and processes that work together to retain resources in a landscape and provide goods and services; see also *biodiversity.*

geomorphic. Pertaining to the shape of natural or constructed landforms.

goal. A desired or planned outcome of landscape restoration.

goods and services. In the context of ecosystem functioning, the provision of shade, shelter, habitat, hydrology, fertility cycling, and microclimate so that environmental conditions facilitate the persistence of biodiversity. This is a narrower definition than found in Daily (1997); see also *ecosystem goods and services.*

gradsect. Transect oriented along a gradient such as water flowing downslope or, if wind is the force of primary concern, downwind; originally defined in Gillison and Brewer (1985).

gully erosion. Form of soil erosion caused by water down-cutting along a pathway or channel, to a depth greater than 300 mm deep.

habitat. The place where individuals of a specified species live because environmental conditions there suit their needs.

herbivore. An animal that feeds on plants.

hillslope. The area that continuously falls, gently to precipitously, between an upper watershed edge and a lower stream bed or floodplain boundary.

hydrology. The study of the dynamics of water including its input by rainfall events, retention and storage in the soil by infiltration processes, and cycling back to the atmosphere by evaporation and transpiration processes.

impact. A disturbance that negatively affects the composition, structure, or functioning of a landscape.

indicator. An easily measured or assessed surrogate for a difficult-to-measure attribute of landscape functionality.

infiltration index. An indicator of the potential capacity of the soil surface in a landscape to absorb water from incident rain and runoff.

interpatch. The area between defined landscape patches where resources are more readily transported away.

invasive species. A nonnative species (usually) that occupies space and utilizes resources in a landscape that would normally be occupied by native species. May disrupt normal ecosystem functions and/or cause extinctions of native species as a result.

landscape ecology. The study of dynamic interactions between the connected ecosystems forming a landscape and the environment, including human activities.

landscape function. How a landscape works as a tightly coupled system of geochemical, biophysical, and socioeconomic processes to regulate the spatial availability and dynamics of resources and to provide goods and services; see also *dysfunctional.*

landscape function analysis (LFA). A monitoring methodology where gradient-oriented transects (gradsects) are stratified into patch/interpatch zones and indicators of landscape functionality are assessed. For example, eleven soil-surface indices are assessed to generate three synthetic indicators: surface stability, infiltration capacity, and nutrient-cycling potential.

landscape health. A subjective assessment of the condition of a landscape, such as being in good, fair, or poor health relative to a particular land use.

landscape organization index (LOI). An indicator of the spatial distribution and size of patches and interpatches in a landscape as measured along LFA grad-sects.

LFA. See *landscape function analysis*.

monitoring. The systematic and repetitive gathering of information about attributes and indicators of landscape components and processes, designed to detect trends in the progress of landscape restoration.

nutrient-cycling index. A soil-surface indicator of the potential for the soil/plant/litter complex in a landscape to provide nutrients for plant growth.

nutrients. See *plant nutrients*.

open-cast mining. See *open-cut mining*.

open-cut mining. A surface mining operation where vegetation and soil are removed to gain access to underlying mineral deposits; also known as open-pit, open-cast, and strip mining.

patch. An area in a landscape such as a hillslope that tends to trap and accumulate resources from upslope (or upwind) open interpatches. Patches also trap resources produced by biota living within the patch.

perturbation. See *disturbance*.

physical soil crust. A layer of densely packed soil particles, often about 1 mm thick, which is caused by rainfall impact onto bare soil and has a very low capacity for water infiltration.

pisolites. Concretions of bauxite resembling a pea in shape and size.

plant nutrients. Chemical entities, both organic and inorganic, which are essential for plant growth, for example, nitrogen, phosphorous, potassium, sulphur, and iron.

practitioner. A person who actively engages with stakeholders to restore landscapes by setting goals, designing solutions, applying treatments, monitoring indicators, and analyzing trends.

process. A dynamic, measureable, time-bound action and reaction in a landscape, for example, infiltration, erosion, photosynthesis, mineralization, growth, and seed dispersal.

query zone. That portion of a patch or interpatch area where soil-surface condition indicators are assessed by LFA methods.

reference site. A site serving as a landscape restoration target or benchmark. Landscape function attributes and indicators are measured on the site and compared with those on rehabilitated sites.

regolith. The layer of loose, heterogeneous material covering an ore body, which in mining is loosely referred to as *spoil* and includes materials such as broken rock, gravel, soil, salts, organics, and biota.

rehabilitation. The process of recovering the functions and restoring the processes that permit ongoing provision of goods and services by damaged ecosystems and landscapes, with respect to reference sites, without intending to fully recover predisturbance species composition and other aspects of biodiversity.

restoration. See *ecological restoration.*

resilience. The capacity of a landscape to persist on its previous functional trajectory after being affected by a disturbance.

resistance. The capacity of a landscape to absorb the effects of disturbances with little or no change in structure and function.

resources. Materials such as water, soil sediments, litter, and seeds that are needed for the full functioning of a landscape.

restoration ecology. The science on which the practice of ecological restoration and rehabilitation are based; provides the concepts, frameworks, and technical information used by practitioners.

revegetation. Actions to establish plants on landscapes being restored with species selection being part of the process.

rill erosion. A form of soil erosion caused by surface flows of water and characterized by channels up to 300 mm deep.

RP. Restoration practitioner. See *practitioner.*

runoff. Overland flow of water occurring during and after rainfall events when soil-surface layers become saturated.

run-on zone. An area in a landscape that accumulates resources carried in runoff; see also *patch.*

scald erosion. A form of soil-surface erosion caused by wind and water that is characterized by the extensive removal of the surface layer of soil (often to a depth exceeding 100 mm) to expose an impermeable clay subsoil; scalds typically occur over a wide area.

scale. The extent or size of the landscape area being rehabilitated (geographic scale) or the time period over which monitoring has been conducted (temporal scale).

seed bank. The viable seeds stored in the soil that are capable of germinating when appropriate conditions occur.

self-sustaining. A landscape that does not need any human intervention or exogenous artificial supply of resources for it to persist in the face of natural stresses and disturbances.

self-thinning. The process applied to populations of similar or even-aged and crowded plants where stresses due to the crowding cause mortality or "thinning" of individuals within the population.

sheet erosion. A form of soil-surface erosion characterized by the removal of thin layers of surface soil (typically a few millimeters) by water over a wide area.

shrink/swell. A property defining those soils that, after being wet, shrink significantly on drying to form deep cracks that fill with rainwater; this deeply wets soil profiles before cracks swell shut.

slake test. A measure of the stability of natural, air-dry soil fragments when gently immersed in water. Stable soil fragments do not rapidly slump or disperse but maintain their cohesion and shape when wet.

slumping. The breakdown or dispersion of a soil fragment when immersed in water; see also *slake test.*

sodicity. A property of soils where sodium makes up 5 percent or more of the soil's cation exchange capacity. Sodic saline soils have high concentrations of sodium chloride and sodic alkaline soils have a high pH because of high concentrations of sodium carbonate.

soil macrofauna. Organisms such as earthworms and termites that live in the soil and provide ecosystem services such as improving the capacity of soils to soak up water; see also *biopores.*

soil-surface assessment (SSA). The procedure whereby the soil surface is assessed by ten visual indicators and also for soil texture; these eleven indicators are combined in various ways to generate three synthetic indices reflecting surface stability, infiltration capacity, and nutrient-cycling potential.

spatial. The role of space in defining species distributions (patterns) and biophysical processes in a landscape. This term is often used in reference to scale; compare *scale.*

spoil. The over-burden materials produced from mining to gain access to the ore body; see also *regolith.*

SSA. See *soil-surface assessment.*

stability index. A soil-surface indicator reflecting the ability of the landscape to withstand the erosive forces of water and wind in producing sediment or dust.

stakeholders. People who have an interest in a landscape restoration project.

stress. A naturally occurring event or process that impacts a landscape as, for example, drought, frost, rainstorm, wind, and flooding.

strip mining. See *open-cut mining.*

sward. A grassland or grassy patch so dense that there are no visible signs of resource transport between or around the grass plants in the landscape.

threshold. The point along a response continuum for an indicator of landscape restoration where, for example, further increases indicate that the landscape is becoming self-sustaining or where decreases indicate a reversion to a more dysfunctional state.

transect log. A record of how an LFA monitoring transect divides a hillslope into patches and interpatches; see also *landscape organization index.*

utilization. The level of grazing of pasture plants, which is usually estimated as a percentage consumption of the total available forage (e.g., 20 percent utilization).

waste rock. The rock within mineral ore bodies that has no commercially valuable content.

Further Reading

We recommend the following books because they provide additional background information to the concepts and principles underlying landscape restoration. We have largely selected books that are also in the Science and Practice of Ecological Restoration Series, published by Island Press (www.islandpress.org) for the Society for Ecological Restoration (www.ser.org). That collection, along with the textbooks listed here, form the foundation of the science of landscape restoration.

Allan, Catherine, and George H. Starkey, eds. *Adaptive Environmental Management: A Practitioner's Guide.* New York: Springer, 2009.

Apfelbaum, Steven I., and Alan W. Haney. *Restoring Ecological Health to Your Land.* Washington, DC: Island Press, 2010.

Aronson, James, Suzanne J. Milton, and James N. Blignaut, eds. *Restoring Natural Capital: Science, Business, and Practice.* Washington, DC: Island Press, 2007.

Bainbridge, David A. *A Guide for Desert and Dryland Restoration: New Hope for Arid Lands.* Washington, DC: Island Press, 2007.

Clewell, Andre F., and James Aronson. *Ecological Restoration: Principles, Values, and Structure of an Emerging Profession.* Washington, DC: Island Press, 2007.

Cramer, Viki A., and Richard J. Hobbs, eds. *Old Fields: Dynamics and Restoration of Abandoned Farmland.* Washington, DC: Island Press, 2007.

Falk, Don A., Margaret A. Palmer, and Joy B. Zedler, eds. *Foundations of Restoration Ecology.* Washington, DC: Island Press, 2006.

France, Robert L., ed. *Handbook of Regenerative Landscape Design.* London, UK: CRC Press, 2007.

Hobbs, Richard J., and Katharine N. Suding, eds. *New Models for Ecosystem Dynamics and Restoration.* Washington, DC: Island Press, 2008.

Lindenmayer, David, and Gene Likens. *Effective Ecological Monitoring.* Melbourne, Australia: CSIRO Publishing, 2010.

Society for Ecological Restoration International Science and Policy Working Group. *The SER International Primer on Ecological Restoration.* Tucson, Arizona: Society for Ecological Restoration International, 2004.

Van Andel, Jelte, and James Aronson, eds. *Restoration Ecology: The New Frontier.* Oxford, UK: Blackwell Science, 2006.

Whisenant, Steven G. *Repairing Damaged Wildlands: A Process-oriented, Landscape-scale Approach.* Cambridge, UK: Cambridge University Press, 1999.

About the Authors

David Tongway is a soil scientist and landscape ecologist who worked for Australia's Commonwealth Scientific and Industrial Research Organisation (CSIRO) for thirty-eight years. David retired in 2003 but continues as an honorary fellow based at CSIRO's Gungahlin Laboratory in the Australian Capital Territory (ACT). He has worked in projects examining landscape degradation and its solution throughout Australia as well as in Indonesia, Kuwait, South Africa, Madagascar, New Guinea, Saudi Arabia, Jordan, and Iran. Although most of his work was in semiarid landscapes, he has recently worked in more mesic ecosystems up to the wet tropics. He has been an associate editor of several international journals and editor of a number of books. In retirement, he has assisted many graduate students at universities in Australia and overseas, endeavoring to continue the traditional role of experienced scientists in mentoring the coming generation. David and his wife, Helen, live in Canberra in the ACT.

John Ludwig is a landscape ecologist whose main interest is in the health of rangeland vegetation. His research experiences include deserts, grasslands, savannas, and forests in Australia, the United States, Mexico, and South Africa. From 1996 to 2006 John served as a theme leader for research on landscape ecology and health as part of the Tropical Savannas Cooperative Research Centre (CRC) based in Darwin, Australia. John's leadership was part of CSIRO's partnership in the Savannas CRC. He worked for CSIRO from 1985 to 2007 and after retiring continues as an honorary fellow based at CSIRO's laboratory in Atherton, Queensland. He also worked in the Biology Department at New Mexico State University in Las Cruces, New Mexico, from 1969 to 1985. Over his forty-year career, John has authored and coauthored numerous scientific publications, including books on landscape ecology and statistical ecology. He also served as an editor for *Austral Ecology* for fifteen years and *Landscape Ecology* for four years. He is currently an editor for *Restoration Ecology*. John resides with his wife, Rosalind Blanche, near Tolga, Queensland.

Index

Acacia, 34, 36, 71, 89, 159
Acid,
    forming spoils, 67, 70
mine-drainage, 167, 171
Adaptive
    landscape restoration, 2, 3–5, 24, 27, 40, 45, 63, 88, 117, 125, 169
    learning loop, 3–5, 72, 81, 92, 102
Aesthetics, 79, 111, 129
Africa, 8, 29, 39, 61, 75–78
Agreements
    land use, 4
    mine site lease, 31
A horizon, 56, 98, 148
Alkalinity
    soil (*see* Soil, alkalinity)
    water (*see* Water, alkaline)
Amelioration treatments, 80, 89, 93, 96, 131
Analysis
    data, 3–5, 21–26
Animals
    domestic, 45, 57, 99
    feral, 50, 57, 97, 99, 101–102, 127, 148
Ants, 67
Aquifer, 67, 70, 80
Arid, 7, 19, 27, 40, 45–46, 56
Artesian basin, 46
Aspect, 142
ASWAT test, 155, 173
Australia, 13, 27, 30, 45–47, 56, 87, 97, 108, 118, 144, 154
Australian Capital Territory, 117–128

Bacteria, 1, 67, 111, 146

Banks and troughs, 20–23, 57, 68, 70, 76–77, 84, 92, 102–103, 141
Basins
    drainage, 46, 68–69, 86
Bats, 120
Bauxite mining, 29–34, 71
B horizon, 5, 56–57, 98, 148
Biological diversity, 7, 13, 27, 107–108, 116, 173
Biological processes, 19–21, 32, 55, 72, 83, 124, 141, 164
Biomass, 12–14, 25, 99, 146
Biophysical processes, 8, 165, 173, 178
Biopores, 37–38, 174, 178
Birds, 7, 39, 111–113, 117, 120, 127–128, 132, 141
Borrow-pit, 49
Brazil, 39
Brushpacks, 103–108, 121, 124–128, 174
Buffel grass, 50, 55

Catchments
    internally draining, 93, 95
    micro-, 41–42, 93, 115
Carbon sequestration, 52
Case studies, 27–61, 63, 118, 140, 152, 163
Cattle, 25, 45–47, 50–51, 88–92
Central Australia, 45–46
Chaining, 100
Channel banks, 154
China, 61, 87
Clay, 69, 80, 155
Claypan, 56
Coal mining, 87–95
Colonizers, 146
Complexity, 7, 71, 127, 140, 157–162

Community Groups, 67, 108, 117–128
Compaction (*see* Soil, compaction)
Conceptual framework (*see* Framework, conceptual)
Contaminants, 67, 70, 75, 83, 90
Continuum, 24–25, 34, 140, 164
Contour banks (*see* Banks and troughs)
Cover
    canopy, 105, 159–160
    grass, 52, 80–81, 90, 101, 160
    plant, 45, 56, 146
Cryptogams, 146
Cycling nutrients, 38, 43, 51, 59, 72, 85, 95, 103, 110, 127, 142, 146–147, 149

Data
    continuous record, 142–143
    field sheets, 129, 141
    spatially referenced, 111, 159
Data analysis, 3–5, 21–26
Data spreadsheets, 139, 142, 145, 148–149, 158–159
Data trends, 20, 63–64, 83, 90, 104
Density
    patch, 23, 157
    shrub, 99–101
    tree, 119
Deep-ripping, 21, 32–34, 40, 49
Desertification (*see* Rangeland desertification)
Dispersion (*see* Soil dispersion)
Disturbance, 4, 7, 19, 24, 39, 47, 58, 70, 100, 111, 140, 148, 151
Ditch, 49
Diversity (*see* Biological diversity)
Donkey, 97
Dragline, 89
Drainage
    acid, 29
    internal, 68–69, 86–87, 93–94
Drainage line
    ephemeral, 137,140, 151–156
    natural, 80, 107
Drought, 46, 55, 90, 108, 122
Dung, 107, 111, 120, 122, 146
Dust pollution, 8, 67–68, 78–80, 83, 157
Dynamic processes, 16, 24
Dysfunctional landscapes, 1, 4, 7, 9, 16, 24, 98,

122, 139–142, 160, 174

Earthen banks (*see* Water-ponding banks)
Earthworms, 67, 83, 109, 178
Ecosystem goods and services, 1, 4, 7, 34, 38, 45, 69, 111, 137, 157, 174
Embankments, 75, 130–133
Engineers, 67–68, 82–83, 130, 132–135
Environmental drivers, 26
Ephemeral
    drainage lines, 137, 140, 151–156
    herbage, 56, 105, 146
    plants, 9–10
Erosion
    caving, 155
    fluting, 155
    gully, 5, 23, 25, 45, 47, 51, 92–93, 102, 151–154, 175
    landslip, 69
    rain splash, 146
    rill, 42, 70–73, 81, 90–94, 98, 102, 111–112, 132–134, 147–148, 151, 177
    scalds, 56–61
    sidewall, 108
    soil, 4, 8, 21, 46, 102, 112, 122, 129, 146, 148
    stream, 109, 129
    surface, 23, 38, 56, 81–82, 98
    tunnel, 89, 92, 155
    wall, 79
Eucalyptus, 30, 34, 36, 71, 159

Farmlands
    cleared, 63–64
    former, 116–128
Fauna, 23, 39, 71, 111, 116, 119, 127–132, 141, 160–162
Feedback
    biological, 33, 38, 55
    physical, 33
    positive, 102, 157
    processes, 12–13
Fertilizer, 22, 33–34, 69, 81, 85, 91, 94, 119, 122
Feral animals (*see* Animals, feral)
Fire
    frequency, 99, 159
    prescribed, 25
    sensitive species, 34, 36, 159

wild, 12, 14, 34, 37, 39
Firewood harvesting, 12
Floodplain, 154, 175
Flow
    lateral, 70
over bank, 70, 152–153
overland, 9, 11, 19, 47, 67–69, 82, 89, 99–103,
    107, 110–116, 124–128, 131–135, 141
resource, 8, 142–144, 147, 153
surface resistance to, 13, 16, 36, 124
Forests, 5, 7, 17, 19, 29, 31, 34, 36–39, 108, 117,
    119, 160
Framework
    conceptual, 1–2, 7–17, 20, 27, 86, 122,
        139–140
    interpretational, 86
    species, 69
France, 61
Frogs, 120
Function-based approach, 1–2, 4–5, 7–17, 18–26
Functional landscapes, 1, 17, 19–26,137,139, 145
Fungi, 1, 9, 55, 67, 70, 80, 83, 111, 146, 173

Gains, 11–17
Game hunting, 12, 102,
Goal setting, 1, 4–5, 8, 19, 22, 31, 35, 39–40, 46,
    56, 66, 79, 81, 88, 92, 99, 102, 108, 119, 130
Goats, 45, 57, 97, 99, 102
Gold mining, 29–44, 65, 75, 141
Gove, 29–39
Government Regulators, 8, 67, 72
Gradsect, 70, 141–144
Graphical approach, 29, 140
Grass
    annual, 101
    cover, 52, 80, 85, 99
    exotic, 3, 50, 159
    native, 100
    perennial, 12, 25, 50, 94, 102, 104–105, 147, 159
    tropical, 33
Grazing
    livestock, 12, 45, 97
    management, 46, 50–51, 98–99, 116, 163
    pressure, 9, 46, 50, 56, 98, 102, 112
    regulation of, 58
Green space, 117, 120
Ground cover, 160

Gully erosion (*see* Erosion, gully)
Gypsum, 89, 93, 131

Habitat, 1, 11, 13, 23, 69, 71, 92, 108, 116–117, 128
Habitat complexity index, 140, 157–162
Hardpan, 56, 98
Heavy metals, 67, 75, 79, 81, 85
Heterogeneity, 66, 84, 98, 141, 157
Horses, 46, 50, 97, 132
Humus, 38, 111

Index
    infiltration, 95, 104, 146, 148–149, 175
    nutrient cycling, 38, 59, 72, 95, 104, 146, 149,
        176
    stability, 38, 95, 104, 146, 149, 178
India, 29
Indicators
    key, 22, 70
    monitoring, 137–162
    trends in, 72
Indigenous people, 31, 34–39
Indonesia, 27, 29, 39–44
Infiltration, 8–11, 21, 23, 37–38, 49, 56–57, 69,
    72, 85, 93, 99, 103–105, 114, 149
Interpatch, 142–150, 175
Invertebrates, 55, 70, 80, 83, 85, 109, 11
Irrigation, 81–82
Israel, 61

Kangaroos, 13, 50, 99, 111, 120–123, 127, 132
Kelian, 40
Key species, 69

Landcare groups, 108, 110–112, 115–116
Landforms, 21–22, 39, 67–68, 80, 95, 130
Landscape dysfunction (*see* Dysfunctional
    landscapes)
Landscape function (*see* Functional landscapes)
Landscape function analysis, 23, 35, 137–162,
    164,175
Landscape processes, 4, 16–17, 20, 22–24, 41,
    46–47, 55, 61, 83, 90, 111, 122, 139–42,
    146, 149, 152
Laser technology, 58, 103
Laterite, 29–32, 39–40
Laws, 4, 19

Leaching, 40, 67, 81, 91
Leaseholders, 66
LFA (*see* Landscape function analysis)
Lichen, 146
Line-intercept method, 142
Lithosols, 88
Litter, 8, 22, 34–38, 70–72, 83, 85, 94, 98, 103, 107, 109–115, 124, 135, 146–148
Liverworts, 146
Livestock grazing, 12, 45, 56, 97, 107, 132, 160

Marra Creek, 47, 56
Mass wasting, 155
Mechanical treatments, 47, 49, 100, 125, 148
Metals
    heavy, 67, 75, 79, 83
Mesh
    fabric, 131
Microbial activity, 11–12
Micro-organisms, 1, 10, 16, 19, 55, 116
Micro-consumers, 1
Mineralisation of nutrients, 11
Mining
    bauxite, 29
    coal, 67
    gold, 39
    hard-rock, 65
    mineral sands, 29
    open-cut, 67
    open-pit, 76
    underground, 78, 88
Monitoring, 4–5, 16–17, 22–25, 35, 42, 50, 59, 69, 72, 80–83, 90, 94, 101, 104, 111, 125, 133, 137–162, 164–165
Mosses, 146
Moths, 120
Mulch, 68, 70, 72–73, 83, 85, 130–135
Mycorrhizae, 146

Native
    animals, 111, 116, 119, 127
    plants, 4, 22, 32–34, 46, 50, 58, 93–94, 101–102, 117, 124, 154
Natural capital, 163–164
Nitrogen, 11, 34–35, 69, 146
    plant available, 52, 54, 90–91, 98, 105, 131
Nurse plants, 100, 102

Nutrients (*see* Soil nutrients)
Nutrient cycling (*see* Cycling, nutrients)

Offtake, 12–15
Organic
    carbon, 16, 41, 52–55, 105, 110
    matter, 1, 8, 33–35, 94, 139
Organisation
    spatial, 139, 141–144
Outputs
    external, 8, 139
Overgrazing, 9, 46, 57, 99, 101
Overland flow, 8, 19
    de-energised, 68, 82, 102, 108, 110, 116, 141
    diffuse, 151
    excessive, 99, 109, 132, 153
    interception of, 127
    regulation of, 111–112, 115, 124–125, 131, 147
    resistance to, 107, 114
Oxidation
    pyrite, 79

Palatable forage, 45, 50–51, 56, 97, 100, 194–106
Pasture
    desirable, 46, 50, 89
    production, 47, 49
Pasturelands, 50, 51, 90–91, 104, 107, 117
Patches, 9, 16, 23, 56, 70, 100, 108, 119, 124–125, 141–144, 145, 150, 157
Patterned vegetation, 98
Pedestal erosion, 81, 101, 122–123, 147–148
Percolation
    deep, 79
Phosphorus
    available soil, 1, 33
Physical processes, 13, 16, 19, 20–22, 32, 69
Physical environment, 1, 7
Plastic flow, 79–80
Plants
    native, 33, 91, 117, 132
    structure, 71, 84, 122, 132, 140, 157–160
Plot-based methods, 157–158
Plotless methods, 157–159
Plumes
    seepage of, 80, 83
Point-centered quarter (PCQ), 158
Pollutants, 67, 88

Pollution
  dust, 78, 80, 132
  off-site, 70
Processes
  basic, 1
  dynamic , 7–9, 16
  feedback, 13, 20
  ineffective , 20
  offtake, 12
  output, 139
  weathering processes, 29
Progress
  evaluating, 3, 20, 24, 55, 70, 101
  tracking, 4, 137, 159
Pulse-and-decline, 8

Query zones, 145–149, 176

Rabbits, 45, 97, 99, 108, 122, 127
Raindrop impact, 20, 68, 99, 132
Rainfall
  amount, 9, 12, 29, 67
  intensity, 8, 41
  simulator, 112–115
Rainforest, 41, 44, 99, 132
Rangeland desertification, 45, 97
Radioactive wastes, 86
Red Natal grass, 3
Reference sites, 4–5, 31, 69–71, 81, 85, 90, 101,
    104–105, 127, 144, 159–160
Regolith, 40, 65–68, 76, 79, 87
Regulations, 4
Regulators
  government, 8, 67, 72
Rehabilitation, 4–5, 20, 32
  active, 160
  landscape, 19, 64,
Repair
  landscape, 1, 7, 17, 31, 45, 46, 52, 56, 98, 103,
    129
  self, 123
Repose
  angle of, 83
Reptiles, 120
Reserves, 9–12, 17, 55
Resource
  capture, 110, 116

  mobile, 141, 144, 147
Respiration rate (*see* Soil respiration rate)
Response curves, 140
Restoration
  design, 5, 122, 153
  goals, 4–5, 8, 19, 27, 31, 56, 101, 110, 137, 163
  principles, 24, 27, 55–56, 59, 65, 110, 122, 140,
    152
  procedures, 3–5, 24, 29, 40, 44, 55, 63, 72, 88,
    125
  technologies, 3–4, 7, 20, 25, 48, 67, 85, 128,
    137, 141, 153, 163
Restoring species, 17
Rill erosion (*see* Erosion, rill)
Ripping (*see* Deep-ripping)
Riplines, 71, 92
Road
  embankment, 129
  map, 3, 163
  verges, 129–131
Rock
  coarse, 130–132
  hard, 65–66
  waste, 21–22, 40. 63, 65–71
Runoff, 8–11, 17, 21, 46, 49, 67, 70, 102,
    107–108, 114–115, 124, 129–132, 153

Salts
  soluble, 67, 88, 116
Saline/Salinity, 71, 91–92
Savannas, 5, 7, 25, 29, 30–34, 81–86, 108, 159
Scalds, 56–61, 147–148
Scenarios, 63–135, 163
Scale
  measurement, 141
  sliding, 24
  spatial, 7, 17, 25, 47, 83, 99, 140–144, 149,
    177–178
  time, 71, 80, 177
Sediment, 47, 51, 67–68, 71–73, 83, 90–91, 107,
    129, 133, 139, 154
  trapping, 103, 124, 127
  transport, 21, 115–115
Seed mix, 34, 89, 91, 93
Seepage, 80–83
  plumes of (*see* plumes)

Self-sustaining
    landscapes, 20, 31–32, 67, 68–70, 127
    vegetation, 127, 130
Self-thinning, 177
Services (*see* Ecosystem goods and services)
Shade, 33, 108–109, 111–112, 175
Sheep, 45, 99, 105, 108, 110–115, 120
Shelter, 11, 34, 108–112, 160
Sheet erosion (*see* Erosion, sheet)
Shrub thickets, 141
Shrubs
    unpalatable, 97–100
Sink-holes, 155
Slake test, 67, 148, 155, 177
Slope, 21–23, 32, 40–42, 45, 58, 70, 73, 88. 93,
    98, 107
Simulation model, 110, 115
Slurry, 75, 79
Socioeconomic issues, 1, 4, 8, 89
Soil
    A horizon, 56, 98, 148
    alkalinity, 89
    ameliorants, 80, 189
    amelioration, 98, 131
    B horizon, 5, 56, 98, 148
    biopores, 37–38, 174
    coherence, 148
    cohesion, 148, 155, 177
    compaction, 21, 32, 34, 37
    friability, 51
    humified layers, 40
    particles, 8, 83, 148, 155
    porosity, 109, 146, 148
    shrink/swell properties, 49–50, 55, 59, 177
    sodic/sodicity, 57, 88–89, 95, 178
    swelling clays (*see* clays, swelling)
Soil biological activity, 32, 55
Soil crusts, 37, 41, 81–82, 110, 146, 148
Soil development, 31–35, 71, 85, 90
Soil dispersion, 48, 67, 90–93, 100–101, 154–155
Soil erosion (*see* Erosion, soil)
Soil fauna, 35, 37–38
Soil formation, 67, 80–82
Soil moisture, 9–12
Soil nutrients, 1, 5, 80, 149
Soil respiration, 55, 85, 105, 109
Soil surface assessments, 142, 145, 178

Soil texture, 149
South America, 29, 78
Space-for-time substitution, 35, 52
Spatial organisation, 139, 142
Species composition, 22, 34, 36, 46, 67, 70, 101,
    122, 140, 158–159
Spoil, 63, 87–94
Spot-spraying, 103, 111
Stakeholders, 3–5, 66, 79, 88–89, 99–102, 157, 178
Stick-raking, 100
Stock watering-points, 50
Subsoil, 49, 57
Sulphuric acid, 79
Surface erosion, 23, 81–82, 105, 123, 146, 147
Surface roughness, 112, 147–148
Surface stability, 38, 59, 61, 72, 85, 94, 101, 127,
    140
Surrogates, 316, 20, 137
Synthetic indices, 38, 140, 149, 175

Tailings, 40, 63, 65, 75–85
Tailings storage facilities, 63, 75–85
Technologies
    biological, 20, 22, 55, 101, 164
    physical, 20–21, 55, 57, 164
Termites, 37–38, 67, 83, 174
Time-marks, 25, 140
Topography, 7
Topsoil, 22, 29, 32–34, 40–41, 66–69, 89
Trails, 117
Trampling, 47, 110, 122, 160
Transfer processes, 8–11, 139
Transpiration, 116, 175
Tree
    belts, 107–110
    canopy, 36, 71, 130, 146, 159
    clearing, 12, 107–108, 115–116
    hollows, 71
    species selection, 110
Trigger events, 8–12, 17, 67, 131
Troughs (*see* Banks and troughs)

United States of America, 39, 61, 97

Vegetation
    growth, 42, 141
    native, 79, 88

patches, 16, 23, 56
perennial, 23, 52, 142, 146–147, 153–154
remnant, 49, 58, 61, 116–118, 124
self-replacing, 32
structure, 71, 122, 137, 157, 159, 163
Viable seeds, 22, 42, 67, 69, 93, 177

Wandering-quarter method, 158
Waste rock, 21, 40, 65–71, 76
Water
alkaline, 92
management, 50
Water-diversion banks, 46, 131–132
Water-ponding banks, 45–61, 103–104, 108, 125
Water retention, 20–21, 45–61, 141
Water spreading, 102–105
Water table
rising, 116

Weathering
rates of, 29, 39, 66–67, 131–134
Weeds, 94, 106, 117, 134
aggressive, 69–70, 111
invasive, 69, 121–125
Weir
leaky, 103, 108
Wetlands, 89–92
Wildfire (*see* Fire, wild)
Wildlife, 11, 13, 92, 108–109, 132–133
corridors, 108
Wind erosion, 80, 83, 141, 147
Wind
resistance to, 83
Woodgreen Station, 45
Woodlands, 5, 29, 82, 88–9-, 92–95, 100,
107–110, 118–120, 144, 160–162
Woody debris, 68–70, 108–115, 124–125, 161